MONKEY TROUBLE

Monkey Trouble

The Scandal of Posthumanism

Christopher Peterson

FORDHAM UNIVERSITY PRESS
New York 2018

Fordham University Press has no responsibility for the persistence or accuracy of URLs for external or third-party Internet websites referred to in this publication and does not guarantee that any content on such websites is, or will remain, accurate or appropriate.

Fordham University Press also publishes its books in a variety of electronic formats. Some content that appears in print may not be available in electronic books.

Visit us online at www.fordhampress.com.

Library of Congress Cataloging-in-Publication Data available online at http://catalog.loc.gov.

Printed in the United States of America

20 19 18   5 4 3 2 1

First edition

CONTENTS

Shh! You'll wake up that monkey.

—JOE GILLIS in *Sunset Boulevard* (1950)

Introduction

No, it was not freedom I wanted. Just a way out; to the right,
to the left, wherever; I made no other demands; even if the way out
should only be a delusion; my demand was small, the delusion
would not be greater. To move, to move on!

—FRANZ KAFKA, "A Report to an Academy"[1]

The human is a source of trouble for posthumanism. Committed to disturbing the opposition between human and nonhuman, posthumanist theory has tended to sideline the human from the scene of its theoretical engagements with otherness. The human has become akin to the "Invisible Gorilla" made famous by psychologists Christopher Chabris and Daniel Simons. Seeking to establish the phenomenon of inattentional blindness, Chabris and Simons instructed participants in a psychological study to watch a video of people passing around a basketball. Many participants failed to notice a chest-pounding, ape-suited human walking through the middle of the scene.[2] For those keen to demonstrate their fidelity to nonhumans, the human has likewise become a conspicuous blind spot.

To be sure, the nonhuman turn has yielded a wealth of critical interventions that have profitably altered the landscape of the humanities. Fostered by loosely federated areas of inquiry such as animal studies, systems theory, actor-network theory, object-oriented ontology, and speculative realism, this turn does not so much name a singular doctrine or movement as it does a broad theoretical reorientation that aims to shift our attention toward a concern for nonhuman alterity. Thanks largely to the insights of contemporary theory, "humanism" designates not only an investment in the human as the locus of rationality and agency, or a rejection of religion and the supernatural as guides for ethical and moral action, or a humanitarian appeal to universal worth and dignity. Humanism also implies an ethico-political hierarchy analogous to other forms of discrimination and exclusion, such as racism and sexism. Humanism *qua* taxonomic hierarchy is thus roughly synonymous with anthropocentrism and "speciesism," a term popularized by Peter Singer in the 1970s.[3] Presaging contemporary critiques of human hubris by several decades, Singer called for the political inclusion of nonhuman animals on the basis of the same liberal-pluralist principles that fostered the civil rights and feminist movements. More recent efforts to include plants and things have sought to expand the sphere of inclusivity even further. Humanism has thus acquired as one of its contemporary connotations a speciesist insistence on the exceptionality of the human at the expense not only of nonhuman animals, but also of countless insensate and inorganic entities.

Human hubris undeniably spawns a general indifference to the myriad of nonhuman beings and entities that share "our" world. This insouciance in turn fosters a number of contemporary ills: factor farming and other forms of animal abuse, global warming, depletion of natural resources, species extinction, and so on. Although we have known since Darwin that we are also apes, and more recent research shows that we share the majority of our DNA with chimpanzees, our "narcissism of minor differences" endures.[4] Could it be that humans need to believe in their exceptionality despite all evidence to the contrary? Consider, for instance, the double valence that the term *Anthropocene* affords. To what extent does embracing this vocabulary concede the deleterious effects that humans have wrought on the environment as well as provide a form of ironic consolation? Perhaps we have utterly screwed up the planet, we tell ourselves, but at least we can take credit for it! The human thus reasserts its power in the same stroke as it reproves itself.

Does what we call the human retain any sense outside the discourse of anthropocentrism? Or is the human indistinguishable from what Giorgio Agamben calls the "anthropological machine," which distinguishes human and animal *within* the human itself? According to Agamben, the modern post-Darwinian version of this machine operates "by animalizing the human."[5] This animalization is but the precursor to our humanization insofar as the machine seeks to isolate our specifically animal attributes as a means of defining what is "essential" to the human. The demarcation of human from animal is far from neutral, bearing all the traces of a sovereign decision that operates according to human self-interest. As Agamben observes, "*homo sapiens* . . . is neither a clearly defined species nor a substance."[6] Drawing from an analysis of Linnaean taxonomy, he observes that Linnaeus initially identified *homo* by appending the ancient adage, *nosce te ipsum*, "know thyself." This expression was later shortened to *homo sapiens*: from the Latin verb *sapere*, "to be sensible or wise." The circularity of this formulation leads Agamben to deduce that "man has no specific identity other than the ability to recognize himself. . . . Man is the animal that must recognize itself as human to be human."[7] The human is irreducible to biology because it is overlain with countless discursive mechanisms through which it reproduces its own image.

Broadly speaking, posthumanist critiques of exceptionalism challenge the presumption that animals lack a number of allegedly unique human capacities, such as language, tool-use, reason, imagination, a temporal sense, and awareness of mortality. Posthumanism has also alerted our attention to how species difference does not exclusively function as a figure for some other category, such as gender, sexuality, race, ethnicity, and class.[8] While the category of species often supplies the metaphorics through which racism and other sociopolitical hierarchies are constituted, species has come to be viewed as worthy of critical analysis in its own right. In the nascent years of animal studies, this attention to species led Steve Baker to ask why animal representations—whether literary, pictorial, or filmic— almost always generate interpretations that reduce the animal to its function as a "transparent signifier" of the human.[9] Baker argued persuasively that the "denial of the animal" has the unfortunate consequence of "ruling out one whole area of potential meanings by assuming that whatever else they may have to do with, the meanings prompted by these representations are *not* to do with animals."[10] More recently Jonathan Burt has argued that animal studies must "bring the animal center stage as the main focus of study, sidestepping the issue of the human-animal boundary, and set this study within the overarching context of human-animal relations—not the overarching context of theorizing about humans."[11] Baker and Burt both recall Walter Benjamin's often-cited remark regarding Kafka: "It is possible to read Kafka's animal stories for quite a while without realising that they are not about human beings at all."[12] Benjamin might be taken here to be expressing a certain posthumanist standpoint *avant la lettre*, urging us to resist allegorical interpretations of Kafka in favor of engaging with animal alterity as such. Yet can and do we ever engage with animals *as such* (literary or real)? Can we utterly dispense with the *as if*, the speculative projections, that we as humans bring along with us? Do we not risk *denying the human—*

or rather the persistence of its phantasm—in our enthusiasm to mark a decisive "turn" toward the nonhuman?

Edmund Husserl observes that the earth does not move insofar as it functions as the ground body for our perception of other astronomical bodies. The earth may not be the center of the universe *as such*, but we experience it *as if* it were.[13] This book argues that the human is likewise quasi-immobile insofar as it conditions all attempts to think what is other to it. An originary "detour" through the human—an irreducible *antehumanism*—renders possible any ethical and political reorientation toward the nonhuman. To claim that the nonhuman turn is *irreducibly humanist* is not to say that it is *exclusively human*, as if animals, plants, and things are simply passive objects to whom we are giving a voice. Antehumanism is *not* the antehuman, or even worse, the archehuman: it does not name an essential humanness that emerges prior to our co-constitution in relation to nonhumans. Antehumanism thus differs from human exceptionalism, which rests on a dialectic of possession and dispossession that jealousy guards human ownership of various self-certified abilities.

Antehumanism also contrasts with anthropomorphism, which is often disparaged for uncritically projecting "human" qualities onto nonhumans. The objection to anthropomorphism presupposes that the characteristics one attributes to nonhumans are proper to the human. The accusation thus fails to recognize how the human self-anthropomorphizes by giving itself this or that capacity whose declared absence among nonhumans performatively delineates the contours of the human. In response to the charge of anthropomorphism, some theorists have eagerly embraced it. Jane Bennett, for instance, sings anthropomorphism's "virtues," while Brian Massumi describes the allegation of anthropomorphism as a "risk" that must be assumed.[14] Massumi is no doubt correct that we can productively identify similarities between human and nonhuman in a manner that avoids the "goo of undifferentiation," yet to promise "not a human politics of the animal, but an integrally animal politics," an "*animo-centrism*" in which the human "loses its a priori dominance," is to sidestep the phantasm of the human.[15] Massumi concedes that starting from the animal point of view is "somewhat arbitrary . . . because the poles of tendential movements are ideal: movements from a starting point that was never occupied, because in point of actual fact there has never been anything other than mixtures in nature."[16] Yet whether we call our departure point human or animal, *we experience it as a starting point*, no matter how arbitrary or false. For her part, Bennett affirms anthropomorphism in a manner that too closely reflects the values of traditional humanism. Her efforts "to give voice to a thing-power," for instance, may seek to deprive humans of our exceptional claim to agency, but the ascription of vitality to nonhuman actants is all-too reminiscent of the self-presencing fiction of human sovereignty, the vigorous self-determination of an "I can" that seeks to stave off the ultimate powerlessness: our vulnerability to finitude.[17] Jeremy Bentham famously asserted that philosophers have been asking the wrong question about nonhumans: at issue is not so much whether animals can speak or reason but whether they can suffer.[18] Homing in on the intransitivity that Bentham's question implies, Jacques Derrida observes that "'Can they suffer?' amounts to asking 'can they not be able?'"[19] Suffering does not belong to the realm of the volitional subject presupposed by the discourse of reason and speech: "Being able to suffer is no longer a power; it is a possibility without power, a possibility of the impossible. Mortality resides there, as the most radical means of thinking the finitude that we share with animals."[20] Suffering involves a condition of exposure and vulnerability, an "*impouvoir*" that interrogates the principle of agency itself rather than simply extend it to nonhumans.[21] Anthropomorphism can impute "human" power to nonhumans only by first *crediting* this power to humans as such. To defend oneself against the charge of anthropomorphism by insisting that nonhumans and humans alike bear a capacity for agency thus does nothing to weaken the exceptionalist, sovereign fantasies through which our conception of the human is irremediably filtered.

Posthuman Error

Human exceptionalism is no doubt a phantasm, but phantasms have a way of persisting. However indefinite and empty, the human attracts a *weak* univocity as soon as we assert its equivocity. How else can we declare the human's fictionality except by presupposing some degree of intelligibility to the identity whose error we have just pronounced? Only on the condition of insinuating this fragile sense to the human is the determination of its indetermination possible.[22] As Michael Naas observes: "the phantasm is not an error to be measured in relation to truth. . . . [It is] not a representation or misrepresentation of the way things are but a projection on the part of a subject . . . of the way one would wish them to be—and, thus, in some sense, the way they become, with all their real, attendant effects."[23] This displacement of the truth/error opposition resonates to some degree with the argument advanced by Dominic Pettman in *Human Error*, where he writes that the human is a case of "mistaken identity—or better yet, the mistake *of* identity."[24] He clarifies that the point is "not to *avoid* mistakes, since this is impossible, but to consciously cultivate more *interesting* ones: mistakes not based on the us-and-them principle of the anthropological machine," errors that are "not structured and limited by fixed taxonomy, by defensive or aggressive sovereignty."[25] Pettman wants to reconceive the relation between human, animal, and machine in terms of a "cybernetic triangle," a "*humanimalchine*" in which all three points of this triad interact equally and thereby deprive the anthropological machine of its privilege in "framing and determining the other two."[26] The cybernetic triangle is no doubt a more interesting error than the most rigid forms of anthropocentrism. Similar to Massumi, however, Pettman fails to appreciate that the human—however misrecognized and misnamed—remains the zero point of our relation to alterity. The phantasm of human exceptionalism cannot be so easily vanquished because its error is also its "truth." The human that declares the fallacy of its own exceptionality can do so only from the position of its phantasmatic centeredness. We turn back even as we turn away; or rather, we never turn away from ourselves precisely so that we can turn away from ourselves.

Pettman's assertion that "*we* are the anthro-machine, and our error is to disavow the machinic part of ourselves as well as the animal aspect" builds on Agamben's claim that the human occupies "a space of exception" that is "perfectly empty."[27] Given this fragility of the human, Agamben draws the lesson that the anthropological machine can be stopped: "To render inoperative the machine that governs our conception of man will therefore mean no longer to seek new—more effective or more authentic—articulations, but rather to show the central emptiness, the hiatus that—within man—separates man and animal, and to risk ourselves in this emptiness: the suspension of the suspension, Shabbat of both animal and man."[28] Does it follow from the recognition of this essential vacuity that the phantasm of human sovereignty can be overcome? How can it be arrested if the power and responsibility for its cessation rests with the human? Given how the human has historically defined itself against animal lack, we ought to remain cautious about laying claim to any power to outright jam the anthropological machine. This force risks becoming yet another self-accredited capacity thanks to which the human redraws the human/non-human distinction through the very movement of its alleged erasure. We find a particularly "strong" version of this line of thinking in the recently consecrated philosophical movement of speculative realism, which revels in stressing that "the world can do without humanity."[29] How does the apparent modesty involved in underscoring our relative insignificance ironically betray its own immodesty insofar as the human grants itself the power to overcome itself?

Speculative realism's anticipated liberation from the human contrasts with the "way out" that Kafka's ape, Red Peter, seeks in "A Report to an Academy." He knows that he cannot free himself entirely from humans, so he chooses vaudeville over the zoo, which he views as "only a new cage with bars."[30] Even if his egress "should only be a delusion," he suggests that the "delusion would not be greater" because his demand is "small."[31] His "way out" is concrete and physical, yet the distinction it marks from total freedom also evokes the philosophical dilemma posed by *correlationism*: Quentin Meillassoux's term for the Kantian position that things-in-themselves are unknowable. Meillassoux wants to escape this "'argument from the circle,'" the claim that there can be no X without its givenness for us humans.[32] As he sees it, "correlationists" (an epithet that encompasses virtually all continental philosophers since Kant) hold that "we are locked up in our representations—conscious, linguistic, historical ones—without any sure means of access to an eternal reality independent of our specific point of view."[33] While the Benjaminian reading of Kafka shares with Meillassoux the aim of displacing the human, the former does not call on us to relinquish the human altogether; rather, Benjamin invites us to abide and stay with animals rather than swiftly seek shelter in the comfortable familiarity of the human. He is certainly correct that Kafka's animal stories are not necessarily about humans, but they are also, for that matter, not necessarily always and only about animals. Red Peter's report, for instance, assumes the form of a testimony that recounts his hominization before an ostensibly human audience. And yet, as the fictional Elizabeth Costello remarks in Coetzee's eponymous collection of short stories, Kafka's first-person monologic narration precludes any external perspective that could verify the speaker's identity.[34] Mirroring the human's own specular self-recognition, Red Peter calls himself an ape. Yet is he really an ape speaking to humans? A human speaking to apes? A human speaking to other humans? Absent any corroborating witnesses, we can only take him at his word.

A similar ambiguity troubles Gabriel von Max's painting *Affen als Kunst-kritiker* (1889; *Monkeys as Art Critics*) (Figure 1). Originally titled *The Ladies' Club*, the work depicts a group of apes viewing a painting that faces away from the human spectator.[35] We see only one side of the tableau's gold frame, to which is attached a label identifying the image as *Tristan und Isolde*. Some of the animals gaze intently at the painting, others indifferently look askance. Yet the central, largest monkey stares directly at the

Figure 1. Gabriel von Max, *Affen als Kunstkritiker* (1889; *Monkeys as Art Critics*). Courtesy of Bayerische Staatsgemäldesammlungen, Munich, Germany.

spectator, her legs open in a manifestly "unladylike" pose. The contrast between her "absent" genitalia and her protruding tongue seems to mock both the masculinist conception of female lack as well as the phallogocentric presumption that nonhumans are deprived of language, as if the painting restores *la langue* to the ostensibly nonspeaking ape. Yet the tongue also mocks the human spectator's ignorance vis-à-vis the unseen painting. The gilded frame constructs a vertical border that reframes the human/animal boundary as a site of loss and absence *for humans*. Shorn of our privileged spectatorial position, we cannot know for certain whom or what the painting depicts. For all we know, the frame may encase a mirror in which the monkeys contemplate their own image, much as human spectators of von Max's painting see themselves reflected in the animal gaze of their nearest evolutionary kindred. The label thus testifies to a human presence that is no more verifiable than the identity of either Red Peter or his auditors.

Whomever or whatever Red Peter is, he begins his report by conceding defeat: he cannot comply with the gentlemen's request because the attainment of his humanness has severed him from his former apehood. The solipsistic circularity of Red Peter's report is not far removed from that of Agamben's mechanism for identifying and recognizing the human, as if the story dramatizes a sort of anthropotheriological machine through which

Red Peter recognizes or misrecognizes himself as . . . well, what exactly? Must we choose between an allegorical and an antiallegorical reading of Kafka? Or might we rather attend to an originary allegoricity by virtue of which the identity of Red Peter remains indeterminable?

This allegoricity is similarly at play in the often-discussed opening of *The Animal That Therefore I Am*. It depicts a scene in which Derrida finds himself being seen naked by his cat: "I must immediately make it clear, the cat I am talking about is a real cat, truly, *believe me*, a little cat. It isn't the figure of a cat. It doesn't silently enter the bedroom as an allegory for all the cats on the earth, the felines that traverse our myths and religions, literature and fables."[36] Stressing this cat's singularity, Derrida voices an ethical comportment toward an irreducible other: "*Believe me*," says Derrida. My ears perk up when I hear him commending us to believe because I believe his understanding of belief is not nearly as straightforward as its invocation here suggests. As he elaborates elsewhere in his work on testimony, belief must not and cannot open onto any absolute certainty:

> "You have to believe me" means "believe me because I tell you to, because I ask it of you," or, equally well, "I promise you to tell the truth and to be faithful to my promise, and I engage myself to be faithful." In this "you have to believe me," the "you have to," which is not theoretical but performative-pragmatic, is as determining as the "believe." At bottom, it is perhaps the only rigorous introduction to the thought of what "to believe" might mean. . . . "What is believing?"—what are we doing when we *believe* (which is to say all the time, and as soon as we enter into relationship with the other).[37]

Derrida suggests that every address to another takes the form of testimony, even if we take no formal oath in front of a judge. Even if we lie or hide the truth, "every utterance implies 'I am telling you the truth; I am telling you what I think; I bear witness in front of you to that to which I bear witness in front of me.'"[38] We are therefore asked to believe him when he insists on his cat's singularity, her resistance to functioning as a synecdoche for "animal" or even "cat." One might believe that this cat's unrepeatable singularity is a fact regardless of whether we believe (in) it. Yet we need not doubt either the existence of this cat or the sincerity of Derrida's belief for us to know (or at least believe) that the assertion of her unrepeatable singularity *as such* cannot arrest her generalizability because "this cat" already begins to subsume the singular being under the general, iterable sign. The *as such* of her singularity is not immune to the *as if* of our specular projections, one sign of which is the traction gained by an Internet legend that has christened Derrida's cat Logos, as if some nonhuman, "viral" force ironically wanted to return language to an animal historically denied it.[39] While Derrida does not identify his feline by name, this appellation cannot prevent her "misidentification" beyond logocentric truth, beyond human volition. In short, I believe that Derrida would not want us to believe him entirely; or rather, he wouldn't want us to convert this belief into a constative truth that would seek to tame "Logos."

As soon as Derrida says "my cat," then, this mineness, as well as its purported nonallegorical reality, are riven by an irreducible allegoricity or iterability that lends this cat to infinite reappropriations.[40] As a consequence of this appropriability, I find myself musing about "my" cat, my cat who died this past Saturday. She was also a real cat, *believe me.* Yet her singular, irreplaceable realness does not stop humans from asking me, "do you think you will get another one?" For Freud, mourning "resolves" thanks to this replaceability of lost love objects, but this means that singularity is always threatened by generality, a violent repeatability that deludes us into thinking that the loss of *this* unique being is only one loss among others.

When I talk about my cat, how do I know that the logos is not playing with me? How do I know if I am talking about her or me? In general, when we talk about animals, even if we forego the general singular "animal" and talk about the singularity of this or that absolutely unique, irreplaceable being, how do we know that our talk does not ultimately go *around* them, following the circuit of our own narcissistic investments? The *mise en abyme* of the aphorism that Derrida exploits in *The Gift of Death* captures precisely this play of difference and sameness: *tout autre est tout autre.*[41] Every other is wholly other. Every other is (the same as) every other. That otherness *appears* to and for us means that its alterity is exposed to and contaminated from the start by our linguistic, cognitive, and specular regimes of intelligibility.

> You have to believe me: I am talking about a *real* cat.
> You have to believe me: I was an ape before I became a human.

At stake in both of these injunctions is precisely the leap of faith that we are called to undertake. The nonhuman turn must necessarily take the form of a testimony in which we bear witness on behalf of those who cannot speak, or at least do not speak the same language. The turn addresses itself to other humans while addressing itself to nonhumans. Which of these addresses is primary? Can we even know? I wager that posthumanist scholars find themselves in a predicament similar to that of Red Peter, even though it goes largely unacknowledged. That is, we report to the academy on the topic of nonhuman alterity in our articles and books without knowing to whom and for whom we register our concern for nonhumans. To the extent that posthumanism leaps over the human, it leaps over the leap of faith on which posthumanist testimony depends. Consider the following observation from Coetzee in an article published in the *Sydney Morning Herald* criticizing industrial animal food processing. Rather than appeal to our conscience or to our better natures, he concludes with this startling observation:

> The campaign of human beings for animal rights is curious in one
> respect: the creatures on whose behalf human beings are acting are
> unaware of what their benefactors are up to and, if they succeed, are
> unlikely to thank them. There is even a sense in which animals do not
> know what is wrong—they do certainly not know what is wrong in
> the same way that humans do. Thus, however close the well-meaning
> benefactor may feel to animals, the animal rights campaign remains a
> human project from beginning to end.[42]

Are we really so certain of our self-proclaimed other-orientedness? The human is both present and absent in posthumanism, but so too are those nonhuman beings on whose behalf we testify: *absent* because they withdraw from us through their oblique inaccessibility, their irreducible alterity; *present* insofar as we *intend* them in the phenomenological sense as beings *for us* whose withdrawal never escapes our narcissistic orbit.

Whataboutism

Recent developments in posthumanist theory are undeniably committed to upping the pluralist ante without end, as if suffering from an acute bout of whataboutism: "You think your ethical obligations terminate with sentient beings? What about plants? What about *rocks*? Who thinks of them?!" We should not presume that the limits of our ethical obligations are fully knowable, that we know in advance who and what are worthy of ethical consideration, or even that we know for certain how and where to draw the distinction between "whom" and "what." Indeed, the scope of our ethical responsibility is infinitely reassessable and renegotiable. Yet contemporary theory is replete with various declarations by fiat of infinite hospitality or infinite inclusivity that do little to advance ethics or politics. According to Timothy Morton:

Nonhumans are also filled with infinite inner space. Some of us are
ready to grant this inner infinity to certain kinds of sentient being.
Some are willing to grant it to all sentient beings. Some are willing to
grant it to all lifeforms (this was my position in *The Ecological Thought*).
And some still further out are willing to grant it to all nonhumans whatso-
ever, no questions asked [my emphasis]. These are the object-oriented
ontologists, in whose number I now find myself. I see no inherent reason why what I called the strange stranger in *The Ecological Thought*
should not apply to any entity whatsoever: fireplaces, the Oort Cloud
at the edge of the Solar System, flamingos and slices of pork rotting in
a garbage can. Since lifeforms are made of nonlife, and since what
counts as a lifeform is very much a performative act down to the DNA
level, I see no big reason not to extend the concept of the strange
stranger to cover all entities.[43]

Morton wants to think human and nonhuman coexistence on the basis of a
kind of burlesque of Levinasian absolute otherness: "Quantum theory
specifies that quanta withdraw from one another, including the quanta
with which we measure them. In other words, quanta really are discrete,
and one mark of this discreteness is the constant (mis)translation of one
quantum by another."[44] Here Morton echoes Graham Harman's claim
that we ought to include inter-object relations in our conception of the
inaccessible, Kantian thing-in-itself: quanta remain unknowable to one
another just as they remain unknowable to humans.

The inclusion of everything may seem like the most ethical gesture possible, but precisely what ethical work does it perform? Is it not simply an
empty gesture that leaps over the "ordeal of the undecidable," thus deferring judgment infinitely by deciding in advance against any discriminatory
judgment?[45] While deconstruction has often been caricatured as promoting the infinite suspension of decisions, this accusation confuses *indecisive-*
ness with undecidability; the latter is the condition of any decision, which
rests on a calculation of what is ultimately incalculable: absolute knowledge
of the rightness or wrongness of one's decision.[46] We must act without the
certitude that our action is just. The declaration of infinite inclusivity, by
contrast, does not constitute an ethical or political act. It simply decides
not to decide. It throws up its hands in frustration, concluding that no
determination of the value or relevance of nonhuman lives or things should
guide our politics and ethics. As Derrida remarks in "On Cosmopolitanism," to affirm an unconditional hospitality independent of the conditional forms they assume in practice is to reduce hospitality to a "pious
and irresponsible desire, without form and without potency."[47] To declare

the arrival *here and now* of an all-inclusive *cosmocracy* is to absolve one from engaging in those discriminatory decisions that are the condition of hospitality as such. An invitation to everything is an invitation to nothing. What sort of investments are revealed by the apparent injunction to be more *other-oriented than thou?* What narcissistic impulses are exposed by this all-too-human desire to outstrip other humans precisely by pledging fidelity to the other-than-human?

Toward a "Weak" Posthumanism

The nonhuman turn has too quickly bypassed the question of what we *believe* we can and do accomplish when we declare the demise of anthropocentrism. This leap toward nonhuman alterity may only appear to be practiced by those "strong" versions of posthumanism that impatiently declare the human's decentering in the present: Morton and Harman's object-oriented ontology, Meillassoux's speculative realism, or Michael Marder's "phytocentrism to come," which echoes the Derridean democracy *à venir*, yet falls prey to the teleology that Derrida resists by suggesting that a plant-centered philosophy might "succeed" where "other de-centerings of the human have failed."[48] Yet even among more mainstream posthumanist theories, those that grasp the difficulties involved in moving "beyond" the human, the decentering of the human is understood as an achievable end. As Matthew Calarco writes:

> The leap from a humanist, anthropocentric (and falsely empty) universal to a truly empty, nonanthropocentric one *is not to be achieved all at once.* In order to understand the necessity for this *transition* and to appreciate the stakes involved therein, it is important first to understand how deeply anthropocentric much of our thinking about animals and other forms of nonhuman life is. . . . *In order for this* [presubjective and postmetaphysical] *thought to be completed*, the 'presubjective' site of relation must be refigured in radically nonanthropocentric terms.[49] (My emphasis)

The teleological impulse here could not be more evident. It lies within our abilities to "complete" our self-posting. Posthumanism just needs to persevere. Calarco patiently awaits posthumanism's achievement, but its advent is no less presentist for all that it is deferred. Its patience is therefore also its impatience.

Calarco's teleological approach is certainly not unique among contemporary posthumanists. In *Before the Law*, for instance, Cary Wolfe justly takes issue with theorists such as Roberto Esposito who efface distinctions between different forms of life. Resisting the normative frameworks that value some lives more than others, Esposito argues that "every form of existence, be it deviant or defective from a more limited point of view, has equal legitimacy for living according to its own possibilities as a whole in the relations in which it is inserted."[50] Wolfe counters that this "principle of unlimited equivalence for every single form of life" amounts to a "cop out" that homogenizes disparate life forms.[51] How could we defend the eradication of viruses and bacteria if all lives should equally be allowed to flourish? Seeking to find the middle ground between unconditional hospitality and immunitary protection—the latter understood both as literal, bodily protection and as the inoculative, exclusionary logics on which all communities are based—Wolfe nevertheless implies that absolute inclusivity lies within the scope of human potential: "Immunitary indemnification is the condition of possibility for any possible affirmation, thus opening the community to its others—potentially, *all* its others. . . . We *must* choose, and by definition we *cannot* choose everyone and everything *at once* [my emphasis]."[52] A few lines later, the necessity of this deferral is reiterated: "All cannot be welcomed, nor all at once."[53] Although he explicitly acknowledges that exclusion cannot be excluded, the "at once" nonetheless implies that conditional hospitality is a temporary problem. No unconditional hospitality, not now. But if not all at once, then incrementally? On the one hand, Wolfe seems to accede to the Kantian regulative idea, stating that "it's not that we shouldn't *strive* for unconditional hospitality and endeavor to be fully responsible."[54] The unconditional thus operates as a guiding, aspirational principle whose reach always exceeds our grasp.[55] On the other hand, Wolfe imagines unconditional hospitality as postponed yet ultimately attainable: we cannot include everyone and everything *at once*, but the promise of unconditional hospitality is bequeathed to a future *present* in which "potentially, *all* its [the community's] others" will be included.

What remains unthought in this analysis is how immunitary indemnification is both a condition of possibility *and impossibility* of unconditional, universal inclusivity. The answer to the question that Wolfe earnestly poses—"Who knows how many others [warrant ethical consideration]?"— must *permanently* withhold itself.[56] To be accountable for previously unidentified others requires that "their" (whose?) final sum remain uncountable. As soon as their totality becomes measurable and knowable, we "know" where the lines of ethical consideration are to be drawn. The advent of unconditional hospitality would be worse than the autoimmune disease it seeks to cure because it would solidify precisely those borders whose permeability and fragility permit us to continually reassess who or what is worthy of ethical consideration.[57] The inestimable others of the cosmocracy to come guarantee that we will have never been able to guarantee the successful inclusion of *all others*.

Teleology also rears its head in the frequency with which posthumanist theorists recount a familiar Freudian fable: the incremental displacement of human narcissism consequent to a series of allegedly "wounding blow[s]": the Copernican blow that unseated the human from its privileged position at the center of the universe; the Darwinian shock that cast humans among the world of animals; and the Freudian one that elevated unconscious mental processes above conscious, intentional, human agency.[58] No doubt Copernicus, Darwin, and Freud each impacted western thought enormously. Yet the rhetorical work performed by Freud's fable of gradual human decentering not only gives us an oversimplified psychohistoriography, but in so doing too easily serves the human narcissism it is meant to diagnose. Dennis Danielson suggests that the Copernican cliché "functions as a self-congratulatory story that materialist modernism recites to itself as a means of displacing its own hubris onto what it likes to call the 'Dark Ages.'"[59] *We moderns are special because we know that we are not special.* As I argue in chapter 3, Meillassoux's charge that Kant's Copernican revolution re-centers the human, and therefore ought to be understood instead as a kind of Ptolemaic reactionism, is belied by medieval cosmology, which held that the earth occupied a position of disgraceful insignificance. The story of the Darwinian wound is likewise overstated because it does not take into account the proto-evolutionist thought of a range of eighteenth-century naturalists and philosophers, such as Linnaeus, Buffon, and Rousseau, each of whom were both captivated and revolted by the physical resemblances they observed between apes and humans.[60] Anxiety paired with fascination toward human/animal affinities was pervasive prior to the *Origin*, thus complicating the received view that evolutionary theory initially attracted only denunciation.[61]

Donna Haraway adds to the three Freudian wounds a fourth, "the informatic or cyborgian, which infolds organic and technological flesh and so melds that Great Divide as well."[62] French ethologist and philosopher Dominque Lestel has argued similarly that scientific investigations into animal language and behavior over the last few decades have inflicted a "fourth narcissistic wound" by demonstrating that nonhuman animals are endowed with at least a weak form of subjectivity, even if they are not necessarily persons or individuals.[63] This accrual of wounds leads Pettman to observe that "humanity is far more robust than Freud thought and can withstand dozens of ego bruises before it admits to being an ex-centric being."[64] Pettman implies here that the affirmation of our eccentricity lies within the horizon of human power, if only humans could resolve their conflicted and ambivalent attitudes toward difference. While Freud argued that the disavowal of difference is never wholly successful, resulting "in two contrary attitudes, of which the defeated weaker one, no less than the other, leads to psychical complications," it seems equally true that no *avowal* could guarantee its success either, unless we believe that the psychic forces of affirmation and negation are subject to human mastery, a doubtful supposition if disavowal names an originary mechanism of psychic defense that allays the threat posed by alterity.[65] The displacement of the human is no doubt essential and urgent, but its decentering does not belong to the dialectic of success or failure, a teleology whose outcome would be either knowable or attainable. Its necessity is also its impossibility. The posthuman thus belongs to the order of the promise, the unforeseeable *arrivant* whose advent necessarily escapes determination. Who other than the utmost humanist human will have been in a position to determine that anthropocentrism's traces have been successfully effaced? By what measure will we have calculated that we have become thoroughly posthuman?

The human who shoulders accumulative narcissistic wounds can pat himself or herself on the back for becoming progressively more other-oriented (self-congratulation in this regard is not all that different from self-flagellation). No doubt the most histrionic versions of this narrative belong to object-oriented ontology and speculative realism, which reject the human-centered world wholesale in order to access what Meillassoux calls "the *great outdoors*."[66] Yet throughout this book I do not oppose something like the "truth" of correlationism to the "delusion," as Red Peter might call it, of the great outdoors. On the contrary, the distinction that we draw between truth on the one hand and fiction, belief, and error on the other will come under pressure by insisting that anthropocentrism constitutes an irreducible belief. The exposure of its falsity can thus only lead to the installation of another belief or set of beliefs: *the "truth" of anthropocentrism's fiction.*

The attention that posthumanism devotes to an ever-widening array of nonhuman beings and entities seems to have been inspired by Lewis Carroll's Walrus: "The time has come . . . to talk of many things:/Of shoes—

and ships—and sealing-wax—/Of cabbages—and kings—."[67] Does setting the cabbage on the same level as the human not risk reaffirming our sovereignty, an anthropomorphic power inscribed by the name we give to this vegetable: "head plant" (from the Latin *caput*)?[68] The critique of human hubris is vital and paramount, but anthropocentrism cannot be displaced through a logic of reversal that elevates immanence above transcendence, horizontality over verticality. The following chapters draw from Husserl's notion of "immanent transcendency," a condition of belonging to the world by not belonging to it, of experiencing alterity from our "exceptional," zero point of perception.[69] Human and nonhuman alterity is both *of* us and *for* us: ontically independent yet phenomenologically dependent. As Dan Zahavi notes, "without asymmetry there would be no intersubjectivity, but merely an undifferentiated collectivity."[70] Cultivating a theory of human/nonhuman relationality that affirms rather than denies the asymmetries spawned by our phantasmatic humanness, I explore literary texts and films that probe the difficulties involved in turning toward the nonhuman. This list of textual "objects" is far less exhaustive than the litanies that suffuse the writings of a number of contemporary theorists, whose interminable catalogues tirelessly strive to say "yes" to everything. Judith Butler once wrote of the "embarrassed 'etc.'" that often closes the litanies of contemporary identity politics ("color, sexuality, ethnicity, class, and ablebodiedness").[71] Yet perhaps we need not be embarrassed by our necessarily partial pluralism, or rather, we ought to bear this embarrassment as one sign of the necessary irreconcilability between pluralism's necessity and its impossibility. I am therefore obliged to include an unavoidably limited number of objects to which I find myself especially drawn. Their collection is arbitrary in the best sense. They are not random Latourian assemblages to which I bear a horizontal relation. Rather, they are objects that I have *arbitrated* as worthy of attention no doubt according to personal investments and cathections of which I am not wholly conscious and which I cannot wholly justify or explain. If I prefer to give up the pretense that these objects and these objects *only* are uniquely suited to the task at hand, then I hope I will be forgiven.

Chapter 1 explores how the human has become something of a scandal for posthumanism: an impediment to "horizontality" and "immanence," those watchwords of the nonhuman turn. While theorists have largely advocated a nonhierarchical and porous conception of the human/nonhuman distinction, I argue that what remains unthought is the *a-porosity* between humans and nonhumans. Our knowledge of both human and nonhuman alterity is fundamentally *aporetic*: "without passage" in the original Greek meaning of this term. A fundamental asymmetry dogs the nonhuman turn at every turn. To explore this asymmetry, I consider the experiments of Irene Pepperberg (parrots) and Herbert Terrace (apes); both set out to trouble the presumption of nonhuman linguistic lack yet ultimately reasserted its absence. Pepperberg, for instance, determined that parrots produce "vocalizations" rather than language. Yet her conclusion depends on an unspoken metalanguage that does take into account the arbitrariness that permits her to fuse the signifier "language" to its allegedly human signified. While both Terrace and Pepperberg remain tethered to a profoundly humanist conception of language, however, I argue that lending the name *language* to animal communication belongs to the same humanist appropriation it aims to escape. "Language" is not simply one signifier among others. It is privileged both because it is self-referential and because humanism defines it as what animals lack. Yet language also shares with all other signs a phantasmatic univocity that interrupts, if only in principle, the sign's intrinsic equivocity. Thanks to this univocity, what we call animal language is always already at least minimally human.

Chapter 2 explores Coetzee's *Foe* and Charles Chesnutt's "The Dumb Witness" through Husserl's notion of analogical appresentation, which holds that our knowledge of others is always indirect and partial. I argue that Susan Barton's apparently well-intended desire to give voice to Friday clings to a familiar political platitude: the presumed value of speech over silence, a silence that she understands as rendering him less than human. That Susan never examines Friday's mouth, moreover, means that he may not truly lack a tongue (as Cruso claims), in which case his silence could be elective rather than violently inflicted. This withheld speech parallels that of Chesnutt's eponymous dumb witness, a slave named Viney who is revealed to have been feigning muteness in order not to reveal the location of her master's deceased father's will. Chesnutt and Coetzee's narratives both stress how the boundary between speech and silence is incessantly crossed, as well as how the "gift" of speech always conceals a sovereign, auto-affective fantasy to hear oneself speak.

Chapter 3 situates Lars von Trier's *Melancholia* (2011) and Alfonso Cuarón's *Gravity* (2013) in the context of speculative realism and object-oriented ontology. Both films foreground the power of nonhuman agencies to frustrate human intentionality. Exploring how the experience of loss attracts both buoyancy (*Gravity*) and ponderance (*Melancholia*), I argue that the noncorrelationist "defiance of gravity," whereby all objects attract the same interest and concern, cannot be sustained. Instead of lending support to OOO's plane of immanence, these films dramatize an immanent transcendency that affirms the human as the inexorable "origin" of the world. I thus develop the notion of an orbital "I" that eschews the false choice between Ptolemaic centeredness and Copernican decenteredness.

Chapter 4 turns to Walt Whitman's *Leaves of Grass*, whose catalogues of nonhuman things have recently attracted the interest of political theorist Jane Bennett. Drawing from a poem in which Whitman instructs us to judge "not as the judge judges but as the sun falling round a helpless thing," Bennett reads Whitman's lists in terms of a magnanimous and capacious ethics and politics that she names *solarity*: an unconditional, indiscriminate hospitality that echoes the cosmocratic plane of immanency endorsed by object-oriented ontology.[72] Despite its purported spaciousness, this hospitality nevertheless depends on an explicit bracketing of the poetic "I," a voice that a number of critics (most infamously D. H. Lawrence) have rebuked for its voracious incorporation and erasure of difference. Bennett's desire to exclude exclusion, moreover, fails to take into account the irresolvable tension between unconditional and conditional hospitality. As Derrida argues in *Rogues*, the borders of democracy are unconditionally traversable in principle, yet subject to all sorts of conditions, decisions, and determinations in practice. This opens democracy up to the dead as well as the living, the inanimate as well as the animate, the insentient as well as the sentient, yet it does not proclaim the full and final advent of absolute inclusivity. While Whitman's "ship of democracy" steers us toward the safe shores of a democracy absolved of its internal conflict between unconditional freedom and conditional equality, I conclude that an all-inclusive cosmocracy always remains to come.

Unlike Red Peter's testimonial, this book reports to the academy in response to no explicit request, under no compulsion, and certainly without any intent to make a monkey out of posthumanism. Peter Gratton observes that the development of speculative realism and object-oriented ontology has spawned a number of impassioned scholarly exchanges, some of which have degenerated into Internet "trolling" as people choose different sides in the debate. As Gratton sees it, "the internet's promise of some vaunted exchange of ideas has given way in too many cases to the exchange of put-downs and pile-ons."[73] Whatever one concludes about the merits of speculative realist theory, a number of major academic presses are publishing it.[74] We would be unwise to ignore its impact, for better or worse.

Moreover, if these theories largely employ a rhetoric that too eagerly dismisses the insights of the last two-hundred years of philosophy, a rhetoric that gleefully throws the correlationist baby out with the humanist bathwater, then we would do well not to replicate its hurried impatience. The blanket rejection of the Kantian critical turn *comes from somewhere*, ironically despite its pretentions to pure immanence, of "subjectless objects," as Levi Bryant puts it.[75] To what extent does the philosophical appeal of subjectless objects emerge—at least in part—in response to contemporary threats to the humanities as a discipline? The conditions that have motivated so many scholars to turn toward nonhuman alterity are no doubt varied and overlapping. Beyond the exigency of resisting human narcissism, a problem that, as I argue throughout this book, cannot be addressed by seeking to inhabit a space of apparent nonnarcissism; beyond posthumanism's sincere ethical commitment to recognizing nonhuman lives *as lives*, we might consider how the nonhuman turn also permits scholars to "manage" or "act out" fears in response to diminishing job prospects and precarious employment. As Gratton observes:

> The loss of numerous teaching jobs in philosophy has happened at the same time as scholars have had easier access to publishing options online, which has meant a heavier burden to identify oneself quickly, to make a name before one even has a name on a regular paycheck. . . . To make it for real these days, the cynical will claim, you must have a system. . . . Better, too, to have that system as soon as possible: Plato gave his philosopher kings some 50 years to develop their chops; he obviously never had to upload his CV to Interfolio. You'll need bona fides for applications for jobs that no longer exist in a discipline funded off, it seems, bake sales and whatever change falls out of the pockets of the Dean of Business.[76]

Susan Sontag once observed that disaster films allow the spectator to "participate in the fantasy of living through one's own death."[77] Has the possibility of human extinction augured by the Anthropocene coalesced in the academic psyche with the changing climate of the humanities (more cold than hot), as if to imagine our foretold death from a position of irreducible survival, irreducible because we cannot imagine the annihilation of the humanities except through a fantasy of survival that permits us to outlive our destruction? As Gratton rightly observes, philosophy has spouted a wave of "self-branding academics pumping out articles and books and pushing new systems of thought."[78] Richard Grusin argues similarly that the propagation of various theoretical "turns" recycles the promise of transformation "as a form of academic branding."[79] Self-branding and obsession with novelty permit the entrepreneurial academic the fantasy of escaping extinction precisely by occupying a central role in the human's decentering. The nonhuman turn is certainly not solely entrepreneurial, yet its various guises form an ineluctable Anthropo*scene*—one that no monkey costume can hide.

CHAPTER I

The Scandal of the Human

Immanent Transcendency and the Question of Animal Language

Man has always been the animal who has no end other than that
which he offers to himself; he has always been the being who
liberates himself—through decision—from determination or
essence. The very concept of man yields, and has always yielded,
a post-humanism. What might be more radical is not a celebration
of the overcoming of man, but a focus on the perpetual, insistent
and demonic return of anthropocentricism.

—CLAIRE COLEBROOK, "Not Symbiosis, Not Now"[1]

According to scholars of the nonhuman turn, the scandal of theory lies in its failure to decenter the human.

The real scandal, however, is that we keep trying.

We can no longer presume our privileged and exceptional status above all other beings, animate or inanimate, sentient or insentient. Nevertheless, our phantasmatic humanness engenders an aporetic relation between us and nonhumans. Theorists who focus on nonhuman entities and agencies are surely not entirely unaware of this aporia. In *Vibrant Matter*, for instance, Jane Bennett worries that her theory of object agency risks "the charge of performative self-contradiction" because it emerges from a human subject.[2] This allegation, she observes, "is not so easy to resist, deflect, or redirect."[3] Yet are resistance, deflection, and redirection our only options? Why should we impatiently "bracket the question of the human," as if it were merely an obstacle on the path toward an ever-greater nonhumanist world?[4] If the elision of the human disavows the fundamental aporia that conditions our "access" to the nonhuman, then should we not abide this scandal rather than attempt to step around it?

That the human is a scandal for the nonhuman humanities must be understood in precise etymological terms. The term *scandal* stems from the Greek *skándalon*: a stumbling block or a trap laid for an enemy.[5] In *The Beast and the Sovereign*, volume 2, Derrida offers a fable of sorts that turns on such an obstacle. Focused on the relation between solitude and world in Robinson Crusoe and Martin Heidegger, Derrida begins by pondering what it means to hear the statement "I am alone" all on its own: that is, in absolute terms, as an expression of an "I" who is "absolved, detached or delivered from all bond, *absolutus*, safe from any bond, exceptional, even sovereign."[6] He observes that "I am alone" always implies alterity "because we're always talking about the world, when we talk about solitude."[7] We are always "alone" together, together alone. From these preliminary reflections, Derrida then invites the reader to imagine strolling along the shore of an island, perhaps similar to the one on which Robinson Crusoe becomes shipwrecked. Suddenly we happen upon a stone, "abandoned or placed deliberately," a rock that we have "tripped over . . . as though it were a stumbling block."[8] Inscribed on this *skándalon* is the following sentence: "The beasts are not alone."[9] From "among ten thousand" possible interpretations of this inscription, which Derrida also asks us to read *alone*, as an aphorism enisled from any larger context, he offers two.[10] The first: "I am a friend of the beasts, there are all over the world friends of the beasts, the beasts are not alone. The beasts must not be alone, long live the struggle for the beasts, the struggle goes on."[11] This initial reading is consonant with the nonhuman turn, posthumanism, animal rights, or any other of the myriad contemporary discourses that declare their concern for nonhuman animals. Yet this affirmation of animal affinity and amity is immediately stymied by an alterative reading that blocks our passage from human to beast: "The beasts are not alone, they do not need us, or else they do not need friends."[12] These textual islands offer entirely divergent conceptions of the abyss between human and nonhuman: one bridgeable, the other unbridgeable. What might it mean to declare that animals do not need our friendship? One can imagine the cynical conclusions to which such a statement might lead. To wit: animals are doing just fine by themselves! Their abandonment to abuse and extermination does not call for human intervention and protection, as if to pervert for the purposes of justifying animal abuse Coetzee's claim discussed in the introduction that they are not aware of our benevolence, that they do not understand the wrongness of our misdeeds committed against them. Of course, Coetzee explicitly rebukes animal slaughter, remarking that the treatment of "any living being like a unit in an industrial process" is "a crime against nature."[13] Clearly what interests him is the claim of proximity and affinity that empathy for non-humans implies: "However *close* the well-meaning benefactor may feel to animals, the animal rights campaign remains a human project from beginning to end."[14] Something frustrates the claim of empathic identification with animals. That "beasts are not alone" thus means that what humans and animals share with one another is precisely the impossibility of sharing the same world. We turn toward this isolated, "worldless" stone (as Heidegger famously put it) only to see reflected back to us our own worldlessness.[15]

The nonhuman turn has advanced largely by eschewing this *skándalon* in favor of affirming a shared world. Asserting that "we have never been human," for instance, Donna Haraway stresses the "multispecies crowd" through which humans and nonhumans are co-constituted: "Partners do not preexist their relating; the partners are precisely what come out of the inter- and intra-relating of fleshly, significant, semiotic-material being."[16] This conception of "worldliness and touch across difference," of "species coshaping one another in layers of reciprocating complexity," explicitly acknowledges human/animal power asymmetries—especially in the context of the dog agility training in which she and her Australian Shepherd dog participate.[17] Yet to claim that "we have never been human" is to downplay the seductive power of human exceptionalism, which cannot be exorcised simply by asserting an immersive companionship with animals.[18] Husserl's notion of "analogical appresentation" is particularly salient to the oblique relationship between human and nonhuman. In the *Cartesian Meditations*, Husserl asserts that "neither the other Ego himself, nor his subjective processes or his appearances themselves, nor anything else belonging to his own essence, becomes given in our experience originally. If it were, if what belongs to the other's own essence were directly accessible, it would be merely a moment of my own essence and ultimately he himself and I myself would be the same."[19] Echoing Husserl, Derrida remarks that, "if the other were not recognized as a transcendental alter *ego*, it would be entirely in the world and not, as ego, the origin of the world."[20] That the ego originates the world does not mean that alterity is simply a product of consciousness in any literal sense; rather, it means that every other with whom I come into contact constitutes another point of origin whose perspective I can never fully inhabit. As Derrida asks, "is not intentionality respect itself?"[21] That intentionality never fully grasps the other means that analogical appresentation names the condition of any ethics of alterity. While Husserl developed his theory to account for intersubjectivity between humans, it is perhaps even more relevant to interspecies relations. Our relation to nonhumans is intentional in the phenomenological sense: a directedness toward alterity that emerges from the human's "sphere of ownness."[22] This sphere is not an absolutely self-enclosed, solipsistic space; rather, self and other inhabit a chiasmic space of intersubjective, "immanent transcendency."[23] Intersubjectivity requires at least a minimal exceptionality by virtue of which my sphere of ownness is never fully accessible to others, and vice versa. Being-with presupposes a quasi-transcendence whereby every subject, human or animal, is "taken outside," as the etymology of exception (*excipere*) implies. We belong to the world by not belonging to it. Haraway's conception of "becoming with" nonhuman others thus ironically risks erasing this alterity precisely by refusing the phantasmatic exceptionality that conditions human/animal becoming.[24]

This asymmetry comes into focus if we follow the logic of Husserl's famous transcendental reduction, which asks us to imagine the possibility of a worldless subject in order to inaugurate a phenomenological attitude toward the world (as opposed to the "natural" attitude that views the world as entirely independent of us). Husserl is not encouraging solipsistic doubt, but is interested rather in how the *attempt* to doubt alters our attitude toward the world. We suspend the world, we put it in parentheses, yet it nevertheless remains.[25] The transcendental reduction thus requires a *provisional* rather than permanent suspension of the world. Only by imagining the possibility of a worldless subject can we come to have an intentional relation to the world. If we merely belonged to the world, if we were entirely *in* the world, then we would have absolutely no relation to it.

Derrida extrapolates from this world-forming characteristic of subjectivity to account for how the world is irredeemably altered by the loss of the other, a deprivation that he links to a line from a poem by Paul Celan: "The world is gone, I must carry you" ("Die Welt ist fort"). Crucially, however, Derrida insists that this loss of the world does not commence with the other's death. As he asks later in this same essay,

> Isn't this retreat of the world, this distancing by which the world
> retreats to the point of the possibility of its annihilation, the most nec-
> essary, the most logical, but also the most insane experience of a tran-
> scendental phenomenology? In the famous paragraph 49 of *Ideas I*,
> doesn't Husserl explain to us . . . that access to the absolute egological
> consciousness, in its purest phenomenological sense, requires that the
> existence of the transcendent world be suspended in a radical *epokhē*?
> . . . In this absolute solitude of the pure *ego*, when the world has

retreated, when "Die Welt ist fort," the *alter ego* that is constituted in the *ego* is no longer accessible in an originary and purely phenomenological intuition. . . . The *alter ego* is constituted only *by analogy*, by *appresentation*, indirectly, inside of me, who then carries it there where there is no longer a transcendent world.[26]

Far from lapsing into an unqualified solipsism, the Husserlian transcendental reduction suspends the external world precisely in order to give the ego over to an alterity that it can no longer know in an immediate way. I can access the other only "indirectly, inside of me." Counter-intuitively, the "annihilation" of the world facilitates a nonoriginary relation to the world, which is to say a relation to an otherness that escapes my grasp. Hence, I transcend the world, but the world also transcends me.

Rather than parenthesize the human, should posthumanism not instead take the "insane" yet necessary step of parenthesizing the nonhuman? Nothing perhaps would seem more politically and ethically indecent in the context of the alleged nonhuman turn than to call for the "annihilation" of nonhumans. Dominic Pettman, for instance, has argued that immanent transcendency amounts to a "double gesture" that cultivates human narcissism.[27] Yet if intentionality as such requires the suspension of alterity, if relating to the world compels us to "doubt" its existence, and thus to carry others in the wake of the world's disappearance, then the "exclusion" of alterity is precisely what the nonhuman turn already performs in its own inchoate fashion. To grasp what it means to say that we should stop *trying* to decenter the human requires that we hear "try" less as "to attempt" than as "to sort" or "to cull," latent meanings derived from the French root *trier*. To try is also to discriminate. We should not try to decenter the human as if its full and final accomplishment were attainable, but we *should* try to decenter the human from the viewpoint of countless trials to come. Their verdicts will always remain subject to appeal because they will always be vulnerable to the accusation of having overlooked someone or something. Hence, the phantasm of anthropocentrism cannot simply be replaced with the truth of its excentricity. *We will have never been posthuman.* The nonhuman turn turns out to have been revolutionary in a way that its advocates likely never intended. As with Husserl's transcendental reduction, we can turn toward the nonhuman world only by first having turned back toward ourselves. The nonhuman turn would thereby name a movement of transformation and return according to the double meaning of "revolution" as both change *and* restoration (as in the premodern definition of revolution as astronomical orbit).

That every turn is always revolutionary in this dual sense is precisely why we should remain skeptical of the rhetoric of "the turn" as such. Despite admitting that academia suffers from "turn fatigue," Richard Grusin attempts to "defend and reclaim" the turn for the nonhumanities.[28] As he sees it, the nonhuman turn is more auspicious than previous shifts in "academic fashion" because it bears the potential to "provoke a fundamental change of circumstances in the humanities in the twenty-first century. . . . A turn is invariably oriented toward the future. Even a turn back is an attempt to turn the future around, to prevent a future that lies ahead."[29] Every academic turn promises change, so what makes the nonhuman turn any more propitious? The answer seems to lie in the twofold temporal and spatial sense of the turn. For Grusin, the *spatial* shift from verticality to horizontality pledges:

> to lose the traditional way of the human, to move aside so that other nonhumans—animate and less animate—can make their way, turn toward movement themselves. I hope that . . . the nonhuman turn . . . might in some small way mark the occasion for a turn of fortune, an intensified concern for the nonhuman that might catalyze a change in our circumstances, a turn for the better not for the worse, in which everyone who wants to participate, human and nonhuman alike, will get their turn.[30]

The nonhuman turn corresponds to a temporal and spatial shift that marks the moment when the human steps away from itself in the hopes of affirming an immanent, nonhierarchical relation to nonhuman others. Yet the conventional, humanist pluralism of this hope could not be more patent. Do we know that nonhumans want to participate? What form might this participation assume? Grusin claims that "affectivity belongs to nonhuman animals as well as to nonhuman plants or inanimate objects, technical or natural," but how exactly will extending our concern to plants and things be "politically liberatory" in the same way that previous "turns toward a concern for gender, race, ethnicity, or class were politically liberatory for groups of humans"?[31] In whose *political* interest do we extend our concern to rocks? Perhaps if we were to turn over the stone that Derrida "discovers" on Crusoe's island we might find inscribed another message: *rocks are not alone.* As with the writing on its obverse, this engraving would comprise both a claim of affinity, even affection, and a claim of solitude, perhaps even a rebuke to all those object-oriented theorists who believe rocks need us to speak on their behalf (can the sub-basaltic speak, anyone?). Perhaps the rock "speaks" only to say "I have no desire to speak, thank you very much. Leave me to my solitary, petrified life."

While it may be in the interest of animals not to be tortured and killed, the same cannot be said of this stone for which political liberation would seem entirely indirect and vicarious: a rebranded version of Kant's claim that "all duties relating to animals, other beings and things have an indirect reference to our duties towards mankind."[32] The Kantian view on animals is notorious, but less often discussed is the place of "inanimate objects" in his conception of indirect duties: "The human impulse to destroy things that can still be used is very immoral. No man ought to damage the beauty of nature; even though he cannot use it, other people may yet be able to do so, and though he has no need to observe such a duty in regard to the thing itself, he does in regard to others. Thus all duties relating to animals, other beings and things have an indirect reference to our duties towards mankind."[33] Theorists of vital materialities, however, often imply that we *do* bear direct responsibilities toward nonliving agencies. Bennett suggests that the conception of politics as exclusively human amounts to "a prejudice against a (nonhuman) multitude."[34] She is no doubt correct that things bear a capacity "not only to impede or block the will and designs of humans but also to act as quasi agents or forces with trajectories, propensities, or tendencies of their own."[35] Yet how precisely does this force of things bear on the problem of political discrimination? Discussing Darwin's treatise on worms, she asks, "can worms be considered members of a public. . . . Are there nonhuman members of a public? What, in sum, are the implications of a (meta)physics of vibrant materiality for political theory?" Bennett draws upon Dewey's conception of a public as an alliance formed in response to "a shared experience of harm."[36] That this formation is not necessarily voluntary or intentional leads her to conceive it as a virtually boundless network of actants, including "dead rats, bottle caps, gadgets, fire, electricity, berries, [and] metal."[37] No doubt a myriad of nonhuman entities bear a capacity to "catalyze a public," but does it follow that they all *experience* the harm around which this public coalesces, in which case we could justifiably call their exclusion *prejudicial*? Is it "wrong to deny vitality to nonhuman bodies, forces, and forms" because they have a "face" in the Levinasian sense, in which case we have a direct responsibility not to harm them?[38] Bennett does not directly address this question. Instead, the political exclusion of worms, dead rats, and bottle caps remains a prejudice in search of a harm. Indeed, she seems to backpedal on the suggestion that these things ought to count as members of a public, conceding that she does not wish to "'horizontalize' the world completely," but rather "to inspire a greater sense of the extent to which all bodies are kin in the sense of inextricably enmeshed in a dense network of relations. And in a knotted world of vibrant matter, to harm one section of the web may very well be to harm oneself."[39] For all the talk of expanding the sphere of politics to include innumerable nonhuman entities, it seems that their "inclusion" ultimately answers to a set of ethical duties that return to the human, as if by doing "harm" to *anything* other one harms oneself. The ethical and political responsiveness that Bennett champions is therefore no less oblique than Kant's insofar as her alleged concern for the vitality of everything is performed on behalf of the human.

Grusin likewise reinscribes the centrality of the human when he suggests that the nonhuman turn must also turn toward "the nonhumanness that is in all of us," by which he means the animal embodiment that humanism has historically disavowed. This turn away from the human thus also *turns back* to it precisely at the place of the *nonhuman within the human*. While the turn as historical shift may be "invariably oriented" toward the future, the turn as spatial metaphor invariably bears within itself an intrinsic *variability* whereby the project of radical immanence turns on itself precisely by turning back to the self. The turn as hope—which Emily Dickinson calls "the thing with feathers," swoops down toward immanence from the human's "transcendent" perch above the nonhuman.[40] The point is not to catch Grusin in a posthumanist "gotcha moment," but rather to ask why theorists remain invested in the rhetoric of "the turn" despite its manifest theoretical inadequacy. If we are unable to abandon the *fiction* of transcendence, then the nonhuman turn amounts to a teleological circle whose goal of absolute inclusivity always pivots around the human, no matter how other-oriented it wants to be. In the final analysis, Grusin's concern for the nonhuman answers to an all-too-human ethico-political imperative. Is it not time to turn away from the turn, to concede that responsiveness to others conceived as a turn always seeks to mask an originary turn toward the self that is the condition of possibility for any ethics of alterity?

The As If As Such

We may genuinely *believe* that we desire to do without the humanist fictions of exceptionalism and transcendence. Haraway is no doubt sincere in this regard when she declares that "human exceptionalism is what companion species cannot abide."[41] Yet what happens to the exceptionalist phantasm once its intolerability is proclaimed? As Michael Naas argues, all phantasms involve an *as if* that attempts to pass as an *as such*: a "speculative fiction" that poses as an "inflexible law."[42] Discussing the centrality of the phantasm in Derrida, Naas writes that "the phenomenon of the phantasm cannot fail to be sustained by the desire, by the temptation, to believe."[43] We must therefore come to terms with:

> the force and tenacity of a phantasm that, metaphysically speaking, does not exist but that we *believe* exists, a phantasm that would be nothing other than our belief in a phenomenon that transcends itself, that spontaneously gives rise to itself—like an Immaculate Conception. For in any consideration of the phantasm one must emphasize less the ontological status of the phantasm than its staying power, its returning power, I would be tempted to say its *regenerative power*. In a word, one must emphasize the fact that the phantasm lives on, the fact that, to cite an English idiom, it seems always to have "legs."[44]

The specific phantasm to which Naas is referring in this passage is the fiction of auto-affection that Derrida put into question in *Voice and Phenomenon*. Whereas Husserl maintains that the closed space of interior monologue constitutes a realm of pure expression in which sign and meaning are aligned insofar as "speaker" and "listener" are identical, Derrida argues that this ostensibly interior world is always exposed to the exterior world of representation, iterability, and difference.[45] The phantasm of auto-affection permits me to believe that my language and meaning is absolutely tied to myself. This phantasm thus denies the "truth" of hetero-affection. Yet Derrida also stresses that auto-affection names "what our desire cannot not be tempted to believe."[46] The French expression *vouloir dire*, "wanting to say," links this desire to meaning. To mean is always to desire an impossible coincidence between sign and meaning, a desire to remain unexposed and inured against all threats to one's integrity and self-presence.[47] The irreducibility of this auto-affective phantasm is precisely what Naas captures by suggesting that it "has legs."

The scandal of human exceptionalism similarly has legs insofar as its staying power belies any simple curative. Absolved of its relation to the human *as* center, the nonhuman turn wants to reveal decenteredness as the human's "truth." If historically the human has presented itself *as if* it were the center, then the posthumanities aim to show us (humans) that we are not truly the center *as such*. Anthropocentrism is no doubt a speculative fiction, but it is also what our desire cannot fail to be tempted into believing. While Derrida does not discuss auto-affection in terms of the human/animal relation in *Voice and Phenomenon*, anthropocentrism depends precisely on the auto-affective fantasy of pure self-coincidence, of an absolutely porous self-relation that nourishes human narcissism: *nosce te ipsum*. Yet this narcissism cannot simply be evaded. As Pleshette DeArmitt argues, "one cannot simply dispense with narcissism, and to attempt to occupy such a position would even be perilous."[48] Indeed, Derrida maintains that absolute non-narcissism extinguishes alterity and thus equates with the worst narcissism possible:

> There is not narcissism and non-narcissism; there are narcissisms that are more or less comprehensive, generous, open, extended. What is called non-narcissism is in general but the economy of a much more welcoming, hospitable narcissism, one that is much more open to the experience of the other. I believe that without a movement of narcissistic reappropriation, the relation to the other would be absolutely destroyed, it would be destroyed in advance. The relation to the other—even if it remains asymmetrical, open, without possible reappropriation—must trace a movement of reappropriation in the image of oneself for love to be possible, for example. Love is narcissistic.[49]

Narcissism is both necessary and impossible. It corresponds to a desire for oneness and sameness whose seductive power is both illusory and unrealizable.

This power underscores a crucial ethical implication of Husserlian analogical appresentation. Husserl employs the term empathy (*Einfühlung*) only sparingly, preferring instead to talk about "the experience of the other" (*Fremderfahrung*).[50] However directly inaccessible and foreign, the other is *experienced* as other rather than merely logically inferred or imaginarily projected. If this experience is always incomplete and indirect, then it follows that ethical duties as such are always mediated as well. For Kant, one should not harm nonhumans because "a person who already displays such cruelty to animals is also no less hardened towards men."[51] The prototype for other humans, however, is oneself. I should be concerned with animal cruelty because it engenders cruelty toward humans, and by extension (or contraction), toward myself. Aside from my ethical concern for animals, my allegedly direct duties to other humans satisfy a narcissistic, self-protective desire to escape harm. That the ethical relation to others—whether human or nonhuman—is always mediated through the self means that the elision of the human commits oneself to an unsustainable, "selfless" ethics of alterity.

What Naas says of phantasms in general is therefore also true of human narcissism. It is not merely an error in need of correction. The "truth" of the human is not only its decenteredness and horizontality in relation to the nonhuman. Its centeredness and verticality do not evaporate simply by wishing it so. These phantasms are *also* its "truth." If anthropocentrism lives under the illusion that it represents the way things truly are, then the posthumanist desire to efface anthropocentrism altogether, to erase any and all of its last vestiges, betrays its own phantasmatic logic, its own *as if* masquerading as an *as such*.[52] Massumi's point discussed in the introduction that both human and animal perspectives are arbitrary starting points because in "*actual fact* [my emphasis] there has never been anything other than mixtures in nature" invokes the as such of immanence *as if* its factuality trumped the as if *as such*.[53] Yet the speculative fiction of the *as if* and the inflexible law of the *as such* are not balanced between mutually exclusive, dueling imperatives. The law of the *as if*, as it were, insinuates itself into the *as such*, and vice versa.

The *"Effanineffable" Name of Language*

We can further trace this co-insinuation of the *as if* and the *as such* in relation to what is perhaps anthropocentrism's most jealously guarded territory: language. A number of language studies have been conducted with parrots and apes since the 1960s (the latter involving the acquisition of American Sign Language [ASL]). The question of whether or not apes possess language has often been reduced to a psycholinguistic problem that measures their communicative capacities against a human definition of what counts as language. Hence, the stabilization of the *meaning of language* is presumed and the phantasm of language's exemplary humanness persists. As Derrida observes in "Eating Well":

> The idea according to which man is the only speaking being, in its traditional form, or in its Heideggerian form, seems to me at once undisplaceable and highly problematic. Of course, if one defines language in such a way that it is reserved for what we call man, then what is there to say? But if one re-inscribes language in a network of possibilities that do not merely encompass it but mark it irreducibly from the inside, everything changes. I am thinking in particular of the mark in general, of the trace, of iterability, of *différance*. These possibilities or necessities, without which there would be no language, are themselves not only human.[54]

The exclusion of nonhuman language is "at once undisplaceable and highly problematic" insofar as its displacement would require a movement beyond the dialectic of "having" and "not-having" language. As soon as humans fix the meaning of language, however, the exclusion of the animal is assured.

Philosopher and ethologist Dominique Lestel has observed that the anthropological dimension of ape language experiments is often treated as if it were merely a form of contamination that interferes with the experimental results. Rather than "subtract" the human element in order to arrive at some kind of "truth" of nonhuman linguistic capacities, Lestel reconceives nonhuman language as precisely what emerges from a process of mutual domestication: a hybrid community that *produces* the talking apes it purports to recognize.[55] Lestel's research marks an important advance insofar as it refuses to bracket the human in hopes of retrieving an inaccessible pure animal language. Yet it remains indebted to a conventional humanist logic that presumes the absence of language among animals prior to what he calls its human *accréditation*.[56] Lestel writes that "it is true that these animals are without language. . . . If they do not talk like human beings do, 'one' speaks for them. The 'talking ape' is acculturated in a special way, because it is integrated into a human community."[57] Lestel does not comment on nonhuman languages outside the context of hybrid, human/animal communities. Primatologists such as Amy S. Pollick and Frans B. M. de Waal have theorized that apes are especially disposed to the acquisition of ASL because gestural communication occurs naturally among them.[58] That humans share with bonobos and chimpanzees (our closest primate relatives) a proclivity for gestures has led scientists to identify this communicative style as the likely foundation of human language evolution. Apes who learn ASL are effectively engaged in an act of *second* language acquisition, in which case it is dubious to claim that they are formerly "without language."

The logic of *accréditation* is also problematic insofar as it reinscribes the human's possessive investment in language. Lestel suggests that the language of apes belongs to "the order of the gift." He continues: "Monkeys do not speak, but researchers can design specific mechanisms through which some great apes can manipulate a kind of symbolic language in their interactions with humans. No chimpanzee has ever spoken like man, but some of them can appropriate segments of symbolic communication with indisputable efficacy. What can one give and to whom? Who can give what to whom? Who can give what to what? And above all: who can give who what?"[59] This law of accreditation not only retains language as what is proper to man, but in so doing ascribes to animals a linguistic poverty that disavows our own inherent linguistic dispossession. As Derrida puts it in *The Monolingualism of the Other*, "I have only one language, yet it is not mine."[60] One inhabits a language that one never fully owns. It comes from the other in the form of an originary "colonial" gesture. Language constitutes a "structure of alienation without alienation," an "inalienable alienation" because its loss does not befall an original possession.[61] Rather, every speech act cites a language that precedes us and therefore attests to an originary linguistic dispersion.

That humans remain divided when it comes to the question of whether animals have language is altogether dependent on an arbitrary decision that fuses the signifier "language" to an identifiable and stable signified. Lacan, for instance, contested Karl von Frisch's discovery of the language of honey bees (for which von Frisch won the Nobel in 1973), claiming that bees rely on *a fixed code rather than fully developed signifiers*.[62] Apparently lost on Lacan was the irony of fixing the signifier "signifier" to a meaning that distinguishes it from the fixity of animal codes. Indeed, he sought to draw a border around language in the hopes of escaping what Paul de Man once described as the vertigo of undecidability: "As anyone who has ever been caught in a revolving door or on a revolving wheel can testify, it is certainly most uncomfortable."[63] The dismissal of undecidability always betrays an effort to extricate oneself from the uncomfortable feeling of remaining within a revolving wheel, of continuously turning around on oneself. We can attempt to exit this revolving door at any time, yet our "decision" regarding the definition of language will not really have decided anything once and for all.

This disavowal of undecidability explains why even the most apparently radical ethological explorations into nonhuman communication prefer to exchange this vertigo for a false stability. In the *Alex Studies*, for instance, ethologist Irene Pepperberg gives an account of her investigations into the linguistic capabilities of an eponymous grey parrot. Pepperberg conducted several experiments with Alex in which he demonstrated an ability to recognize difference and sameness, presence and absence, as well as a capacity to communicate intentionally rather than engage in mere mimicry. For instance, he learned to apply the word *key* to keys of varying color, thus demonstrating an advanced cognitive ability to transfer skills from familiar to novel situations. He also learned to say "I'm sorry" in appropriate contexts: once after he chewed up a grant proposal, another in response to Pepperberg's visible frustration when he refused to cooperate with a routine skills test, and yet another after he knocked a plastic cup onto the floor. Pepperberg acknowledges that she cannot prove that "I'm sorry" constitutes an expression of true remorse (of course, the presence of genuine contrition is equally immeasurable in humans!), but she nevertheless interprets his words as an effective means to defuse a tense situation. Pepperberg designates such communication "peri-referentiality," which she distinguishes from "fully referential" language.[64] An ability that parrots share with the great apes, peri-referentiality employs a symbol as a "mental representation of an item," but it stops short of "full abstract use of a symbol" in which one is able "to talk about qualities of the item, to talk about how you think about the item—the referent—in its absence, to talk about it in future and past tense—but not simply in the sense of a request for something not yet present."[65]

In her memoir, *Alex and Me*, which Pepperberg wrote in the wake of the parrot's death in 2007, she reveals that the name Alex was originally intended as an acronym for Avian Language Experiment, but due to resistance within the scientific community, she revised the acronym to denote Avian Learning Experiment. Pepperberg began her research in the late 1970s at a time when ape language studies were under attack.[66] In 1979, Herbert Terrace, who had previously been a staunch advocate of such studies, published a paper called "Can an Ape Create a Sentence?" in which he claimed that his experiments with an ape named Nim Chimpsky (after the famous linguist) had failed to demonstrate linguistic capabilities in apes. According to Terrace, "the function of the symbols of an ape's vocabulary appears to be not so much to identify things or to convey information . . . as it is to satisfy a demand that it use that symbol in order to obtain some reward."[67] Nim may have signed "banana" when he wanted one, but he did not demonstrate any conception of grammar. Of course, Terrace begs the question as to why grammatical ability ought to mark the gateway to language. If Nim could sign "banana," then he no doubt demonstrated a basic grasp of referentiality. Terrace's article helped fuel an openly hostile attitude toward studies of nonhuman communication, and thus Pepperberg felt compelled to cease claiming that parrots employ language or words, and stated instead that they use "labels" and "vocalizations." As Pepperberg remarked in an interview in 1999: "I avoid the language issue. . . . What little syntax he [Alex] has is very simplistic. Language is what you and I are doing, an incredibly complex form of communication."[68] We should certainly sympathize with the plight of a young scholar trying to gain a foothold in academic research and publishing—which, for all its professed obsession with innovation, is often ironically unreceptive to work that refuses to parrot the status quo—yet we must nevertheless ask how far ethological research into nonhuman languages has actually progressed if a well-established scholar still feels induced to such caginess some thirty years later.

Rather than avoid the question of language altogether, should we not insist instead that the question of where we draw the line between human language and nonhuman "vocalizations" is among the most pressing questions, a question that calls on us with considerable urgency, but that can only remain unanswerable, a question that must be posed and reposed precisely so that it remain open, so that no final definition of language can be imposed? It seems commonsensical for Pepperberg to believe that scientific investigations into the question of nonhuman communication must first come to a decision about what constitutes language itself: how can we determine if animals have "it" if we do not first define what "it" is? Yet any language that seeks to limit the meaning of language can do so only by posing as a metalanguage, which is to say a language that masquerades as not language, a language that has much to say about what language is and how animals do not have it but has nothing to say about the language that authorizes itself to assert this lack. The human thus presumes that it can stand above and beyond language, that it bears the power to delimit the meaning of "language." As loquacious as it is when it denies animals a capacity for speech, the language of science suddenly becomes dumbstruck when it comes to justifying the human's exceptional claim to language. When Pepperberg claims that "language is what you and I are doing," the circularity of her assertion is unmistakable: she gives herself the sovereign power to decide what does and does not count as language, a power that depends on nothing more than what Derrida calls (drawing from La Fontaine) "the reason of the strongest," a power that exempts one from the duty to provide rationales.[69] Parrots thus always fall short of some imaginary threshold. Animals that connect a particular sign with a particular object may display a capacity for association, she suggests, but this is "merely the *first step* toward referential labeling [my emphasis]," which requires "communicative intent."[70] Likewise, the conceptual understandings inherent in peri-referential labeling are "defining characteristics *leading up to* referential communication [my emphasis]."[71] Animal "vocalization" is in the vicinity or neighborhood of language, but is not quite there yet; it is always on the way toward language, but never fully arrives there. Although she suggests that "lack of evidence for truly referential communication in animals is most likely a consequence of our own incompetence," implying that "true" referentiality in nonhuman animals may yet be discovered, this statement is difficult to square with her equation of language with what she deems "truly" complex forms of communication.[72]

Pepperberg may openly reject the view espoused by some ethologists that language "will continue to be redefined by linguists as whatever animals cannot be shown to do," but by employing a "human standard for the term 'referential'" has she not already imposed an untraversable boundary between human language and animal vocalization?[73] Lestel remarks that the criterion employed by ape researchers closely resembles that of the Turing test, which the latter proposed to determine whether a computer can be regarded as intelligent. Just as a computer passes the test if it can deceive humans into mistaking it for another human, "the ape will be recognized as speaking when it makes impossible, for a human being, the distinction between a human being and a chimpanzee, with regard to language."[74] According to such standards, animals would be forever excluded from language. Only if we reduce human language to a code of natural or naturalized signs can it be understood as fully referential. Yet even were we to concede that animal symbolization is less developed than that of humans, this concession would still not justify restricting the term *language* to human forms of communication. That the peri-referentiality of parrots is prescriptive rather than descriptive means that they are effectively barred from language by an infinitely receding horizon that they have no hope of transcending.

It would be all too easy and expedient to brand as "humanist" any argument that reserves language exclusively for humans. If antehumanism is irreducible, however, then any inquiry into animal languages presupposes at least a provisional definition of language that limits our capacity to grasp nonhuman language "on its own terms." Is it even meaningful to suggest that the *as such* of nonhuman language is accessible to us beyond the *as if* of our imaginary projections? On the one hand, we can and should challenge human exceptionalism by showing that language is not the sole province of the human. On the other hand, what we *name* language presupposes what Derrida characterized as early as his *Introduction to Husserl's "Origin of Geometry"* as a "minimal linguistic transparency."[75] We can expand our conception of language to embrace innumerable forms of nonhuman communication, to include the general structure of the trace, and so on, but this expansion requires that we proceed *as if* language *qua* signifier is minimally univocal. Derrida illustrates the interdependency of the univocal and the equivocal in a well-known passage from the *Introduction* that discusses Joyce. Although Joyce sought to unearth "the greatest potential for buried, accumulated, and interwoven intentions within each linguistic atom," this excavation "could only succeed by allotting its share to univocity, whether it might draw from a given univocity or try to produce another. Otherwise, the very text of its repetition would have been unintelligible; at least it would have remained so forever and for everyone."[76] For all its equivocation, Joyce's *Ulysses* is no less exempt from a sort of transitory interruption of the sign's intrinsic heterogeneity.

Even if we accept that signification is subject to an enduring undecidability, this absence of determinable meaning is no more tolerable than finding oneself trapped in a revolving door. We thus "decide" meaning in a number of interpretative contexts in order to mitigate temporarily this vertigo. The signified of "language" may never present itself; it is always *to come*, but we proceed here and now *as if* its meaning has already arrived. And this *as if* is no less evident than when we reject the humanist exclusion of the animal from the domain of language. As soon as we say "language cannot be defined," or "language is not exclusively human," we have already taken a step toward defining language; the intelligibility of these assertions posits some degree of reference, some provisional definition of language, no matter how fallible and precarious, no matter how open to revision and contestation.

The rejection of exceptionalism cannot extricate itself from this tension between univocity and equivocity, between absolute translatability and absolute untranslatability. Language is equivocal *as such* but we cannot avoid believing *as if* it were minimally univocal. The human who wields the Adamic power to name what the nonhuman can no longer be said to lack— in a word, language—thus *reinscribes* exceptionalism precisely through the inclusive gesture that "gives" language to nonhumans. The "gift" of language to animals imposes a univocity that recalls Derrida's *animot*: a homonym with *animaux* that stresses how the catchall "animal" captures the plurality of animals in its linguistic cage.[77] Are we not reinscribing the general singular "animal" despite all the different languages that may exist among nonhuman animals, languages that nonhumans would certainly not call language, but which might go by other names? Does language not constitute a *humanimot* that bespeaks the human's monolingualism? Perhaps my cat has a secret name for her purrs and meows, an "Effanineffable/ Deep and inscrutable singular Name."[78] Hence, the most generous, posthumanist gesture that would lend the name language to her voice must reckon with the antehumanist phantasm that this gesture evokes.

This *humanimot* of language is precisely what Massumi disavows when he claims that "human language is essentially animal."[79] Massumi reads animal play as a prototypical form of "metacommunication," a "simple code" that "produces the conditions of human language."[80] Focused on the ludic gestures of wolf cubs whose play fighting communicates "this is not a bite"—thus marking a distinction between real and figurative combat that is inherently communicative—he asks why, if animal play is protolinguistic, then do we "not consider human language a reprise of animal play, raised to a higher power? Or say that it is actually in language that the human reaches its highest degree of animality?"[81] While he is critical of the "monopoly" that humans claim over language, *his own language* maintains a hierarchical distinction between human and nonhuman by describing animal play as "metacommunicative" and the human capacity for figuration as "metalinguistic."[82] Similar to Pepperberg, Massumi describes animal communication as "language-like," "language *avant la lettre*."[83] Rather than "give" animals language, he takes language away from the human, or rather, "demotes" language to the level of the animal. Yet this demotion is only apparent insofar as human language sublates its essential animality and raises it up to another level. Indeed, this approach seems *even more* conventionally humanist than that of Pepperberg because it remains silent on the sovereign decision that distinguishes animal communication from human language. As I will show in chapter 2, sovereignty in its purest (impossible) form requires silence, lest sovereignty undermine itself by speaking, by providing reasons—in this case, by supplying a rationale for deciding on the distinction between animal communication and human language, as if this decision were not dependent precisely on the *fiction* of transcendence whose perpendicular distance abstracts the subject from the scene. Massumi wants to claim that human and animal difference can be affirmed only in "absolute survey," that is, "without attributing any foundational status" to such distinctions.[84] Absolute survey, which Massumi also calls "immanent survey," describes a perspective that claims not to stand apart from or above what it perceives, as if from a bird's-eye view.[85] Such a transcendent perspective is no doubt a phantasm, but Massumi proceeds as if its error can simply be opposed to the truth of immanence. Absolute survey wants to absolve itself from complicity with the phantasm of transcendence. Seeking to distance itself from this phantasm, however, it maintains its own bird's-eye view "above" the knotty aporia of immanent transcendence.

The march toward immanence that claims to relinquish "our inveterate vanity," to renounce and thereby move beyond our auto-affective, narcissistic fantasies, can only disavow what Derrida calls "the law of the island and the law of the wheel" by virtue of which "my last footstep always might coincide with my first."[86] Discussing the anxiety that Robinson Crusoe experiences when he cannot determine if a footprint he discovers in the sand belongs to a "cannibal," another castaway, or to himself, Derrida suggests that the uncanniness engendered by the footprint's uncertain origin resonates with the unease that Robinson also experiences when he hears his parrot, Poll, say "Robin, Robin, Robin Crusoe, poor Robin Crusoe, where are you, Robin Crusoe? Where are you? Where have you been?"[87] Having found himself temporarily stranded at sea when he explores the opposite side of the island, Crusoe finds Poll's words uncanny because they seem to be an expression of longing and mourning, as if Poll has flown in search of his lost human companion. Robinson believes that Poll has merely learned these words by rote when the former apparently uttered them in an expression of profound isolation and disorientation. Yet the question of who owns Poll's linguistic traces is no more answerable than the question of who left the footprint in the sand. Who's to say that Poll has not reappropriated his master's words for his own purposes? Robinson's linguistic disorientation thus mimes his geographical disorientation: "These are always the two risks of a proceeding [*démarche*]: wander and get lost, or get closed in by retracing one's steps. And that is the Robinsonian trouble with the island. Not get lost and not get closed into the aporia, not get paralyzed."[88] As with de Man's revolving door of undecidability or the uncertainty that Robinson faces when he is unable to determine if the footprint he discovers belongs to him or to another, the apparently straightforward path that would lead us to nonhuman language always poses the risk of going in circles and retracing our steps. In this sense, Lestel is both absolutely correct and absolutely incorrect to say that apes do not have language. Ethologists such as de Waal do not teach apes to learn ASL, but they nevertheless "give" them language precisely by certifying their gestures as language. *Accréditation* therefore takes place as soon as one asserts that nonhumans possess language, even if the language that one "gives" them is already "their" own.

Whence the following paradox emerges: language is only and always human; language is never solely human. Humans have only one language, yet it is not our own.

The full and final displacement of the humanist orientation in and of language does not fall within the scope of human potentiality, within the horizon of what the human can achieve or accomplish. This failure attests to a certain nonpower, a not-being-able, a vulnerability and subjection to language that undermines any effort to secure and delineate the scope of its signification. The most obstinate humanists among us will no doubt continue to insist that language is the inalienable property of the human, notwithstanding scientific and ethological efforts to demonstrate the contrary. Yet only those who would thereby assert with equal confidence that they know what language is could claim to know that animals possess it. One would never be able to *prove* that animals have language, except by summoning forth the same metalanguage that will have authorized their silencing.

Lestel relates an anecdote about an exasperated linguist who asserted at a colloquium on the topic of language among bonobo apes that "there will always be a fundamental difference between bonobos and humans; it will always be humans who organize symposiums on bonobos and not the reverse."[89] The unlikelihood of an ape presenting a scholarly paper to his or her peers on the question of human language does not demonstrate that apes lack symbolic language, but it does underscore a fundamental asymmetry between human and nonhuman animals. Openness to animal alterity cannot neutralize the quasi-solipsism by virtue of which access to the other must be given indirectly precisely in order that the other remain other. That the "gift" of animal language cannot escape the monological sovereignty that it opposes means that we cannot finally distinguish the return of language from its imposition, the gift of language from its theft. If animal language always betrays the stain of the human, then we can vainly attempt to escape this trace (as Crusoe flees the footprint) or endure its uncertain origins. This monolanguage that is not of the human, this singular print that does not belong to us, nevertheless leaves its mark on those animals whom we summon forth to speak from the margins of an enforced mutism.

CHAPTER 2

Sovereign Silence
The Desire for Answering Speech

The Sirens . . . have an even more terrible weapon
than their song—namely, their silence.

—FRANZ KAFKA, "The Silence of the Sirens"[1]

J. M. Coetzee's *Foe* revolves around an alleged absence. According to Cruso, Friday cannot speak by force of a cruel slave master having cut out his tongue. Its loss recalls von Max's painting *Monkeys as Art Critics* that I analyzed in the introduction. Whereas the tongue of the painting's central female monkey pokes out at the viewer as if to satirize the humanist and masculinist presuppositions of feminine and animal lack, this deprivation is thoroughly racialized in the case of Friday. Von Max's painting is not, by itself, racially suggestive, but one cannot read it alongside *Foe* without recalling the long racist history of associating black people with apes.[2]

Foe retells Defoe's *Robinson Crusoe* from the perspective of a woman named Susan Barton who finds herself shipwrecked on the same island with Friday and Cruso (spelled sans "e"). They are eventually rescued, but Cruso dies en route to England, where Susan tracks down the novelist Daniel Foe in hopes that he will help her write a manuscript recounting her adventures. In addition to chronicling her experiences on the island, Susan attempts to restore language to Friday. After initially trying to teach him the names of everyday objects, and later engaging in a failed musical duet using two of Foe's flutes, she finally settles on the idea of teaching him to write. She quickly becomes frustrated with her pupil's lack of progress, as he only seems capable of threading together an incoherent string of letters as well as figures that resemble "row upon row of eyes upon feet: walking eyes."[3]

Susan's motivation to communicate with Friday is not as straightforward as it might first appear. Her apparently other-oriented gift of language both conceals and reveals what she elsewhere terms *her* "desire for answering speech," which she likens to "the desire for the embrace of, the embrace by, another being."[4] In other words, her ostensibly altruistic reasons for restoring Friday's voice obscure her own desire for language and its promise of intersubjective immediacy. Would Friday want to speak even if he could? Moreover, how can we be absolutely certain that he is physically incapable of speech? As Lewis MacLeod observes, the novel provides no definitive evidence that Friday is lacking a tongue.[5] As with Susan, we can only take Cruso's word for it. Derek Attridge also remarks upon the lack of conclusive evidence that Friday's tongue is absent. Even so, he focuses almost exclusively on silence as a product of dominant discourses. Aligning the colonial violence of Friday's silencing, the struggles over authorial voice that arise from Susan's desire to have her story told, and the larger historical processes of literary canonization, Attridge writes, "all canons rest on exclusion; the voice they give to some can be heard by virtue of the silence they impose on others."[6] Yet literary texts do not "speak" precisely in the same manner as living subjects. The legitimation of certain texts at the expense of others is not strictly analogous to Friday's silence, and not only because we cannot rule out the possibility that it is willfully performed rather than repressively imposed. Unlike the metaphorically silenced literary text, a speechless *living* subject does not necessarily dwell in silence.

The penultimate chapter of the novel resolves on the weakly optimistic note that Susan's efforts might finally pay off. After taking a brief respite from their lessons, she returns to discover Friday seated at a table wearing Foe's robes and wig, busily smudging the papers with sequences of the letter *o*. "It is a beginning," says Foe. "Tomorrow you must teach him *a*."[7] The fourth and final section of the novel immediately follows, consisting of two short fragments narrated by a first-person voice whose identity remains undisclosed. The narration begins by repeating the opening line of section three, which was originally told in the past tense from Susan's perspective but is now given in the present tense: "The staircase is dark and mean."[8] As if to foster the impression that Susan and this voice bear a common perspective, the narrator eventually descends into the water in the same area where Susan observed Friday casting flower petals, a mysterious ritual that she had earlier speculated marked the watery grave of a friend or family member who drowned in the wreck. That this "I" describes Susan, Friday, and Foe as now dead means, however, that none of them can inhabit the grammatical position of narrator.

This unidentified "I" thus floats among the wreckage of the sunken ship, never anchoring itself to any referential ground, as if conforming to the dream logic of Freudian "condensation" whereby more than one experience and identity are combined and manifest in a single dream image.[9] Yet we are nevertheless bidden to loosely tether this voice to Friday insofar as his linguistic abilities and the identity of the final section's narrator are equally the novel's most conspicuous and most unfathomable unknowns. Although the source of the narration is not finally knowable, the impenetrability of Friday that preoccupies Susan invites us to associate this voice with him as much as with her. Whereas Friday is a body without a voice, the narrator is a voice without an identifiable body. Not by accident does the final section of the novel follow immediately after Friday learns to write "rows and rows of the letter *o* tightly packed together."[10] The letter *o* visually marks a hole or opening through which emerges a new voice that is not strictly identified with Friday, but which nevertheless draws us inward, giving us to think that the novel might finally fasten the buttonhole that Susan invoked earlier as a figure for Friday's nonexistent tongue.

If Friday's *o* inscribes both an opening and an absence, it also graphically closes in on itself. Like the buttonhole that figures the tongueless mouth, the *o* inscribes an aporia that both opens and closes, reveals and conceals. The reader thus slips through this orificial *o* into a series of enclosed spaces, beginning with the narrator's ascent along a dark staircase in Foe's house that leads to a dim, oxygen-deprived room ("my matches will not strike"). Here we find the dead bodies of Susan and Foe exhibiting signs of advanced decay, and a supine Friday harboring only a faint pulse.[11] The second half of the section abruptly shifts location to the submerged wreck of Cruso's ship, where the narrator finds Friday half buried in the sand and seemingly deceased, and the dead bodies of Susan and Foe appear bloated from prolonged submersion. "What is this ship?" the narrator asks Friday. Grasping the futility of eliciting a response, the voice continues: "But this is not a place of words. Each syllable, as it comes out, is caught and filled with water and diffused. This is a place where bodies are their own signs. It is the home of Friday."[12] What does it mean that we have arrived at the home of Friday? And what is the status of this narrative voice that leads us to the humanly uninhabitable bottom of the ocean floor, a place where language is thoroughly inoperable, a place that cannot properly be narrated by any living, breathing human "I"?

Susan's preoccupation with giving voice to Friday is consistently framed in terms of a language of penetration that would access his hidden interiority: through the eye, mouth, or ear. When Susan recounts having witnessed Friday scattering the flower petals, Foe surmises that it may in fact have been a slave ship rather than a merchantman (despite Cruso's claim to the contrary), in which case the boat would have marked the burial place of "hundreds of his fellow-slaves—or their skeletons—still chained in the wreck, the gay little fish (that you spoke of) flitting through their eye-sockets." Foe continues: "Friday rows his log of wood across the surface of the dark pupil—or the dead socket—of an eye staring up at him from the floor of the sea. He rows across it and is safe. To us he leaves the task of descending into that eye."[13] Whereas Foe employs the image of an empty eye socket in order to figure his descent into the mind of Friday, Susan invokes a different orifice: "It is for us to open Friday's mouth and hear what it holds: silence, perhaps, or a roar, like the roar of a seashell held to the ear."[14] When Susan hears that Friday's tongue has been removed, she develops an absolute aversion to Friday's mouth, refusing to examine its dark recesses when Cruso commands her to do so. Yet the mouth as both an anatomical and *figurative* void is cited frequently throughout the novel, culminating in the final two paragraphs:

> He turns and turns till he lies at full length, his face to my face. The skin is tight across his bones, his lips are drawn back. I pass a fingernail across his teeth, trying to find a way in.
>
> His mouth opens. From inside him comes a slow stream, without breath, without interruption. It flows up through his body and out upon me; it passes through the cabin, through the wreck; washing the cliffs and shores of the island, it runs northward and southward to the ends of the earth. Soft and cold, dark and unending, it beats against my eyelids, against the skin of my face.[15]

Eye sockets, mouths, ears, dark suffocating rooms, deep-sea shipwrecks— these spaces are all aligned with the hidden depths of Friday that Susan longs to plumb. To access this interiority, she must give Friday a voice. Yet her many speeches on the value of speech read like a litany of metaphysical, humanist, and political platitudes—all centered on the indubitable value of speech over silence:

> If the company of brutes had been enough for me, I might have lived most happily on my island. But who, accustomed to the fullness of

human speech, can be content with caws and chirps and screeches, and
the barking of seals, and the moan of the wind?[16]

To live in silence is to live like the whales, great castles of flesh
floating leagues apart one from another, or like the spiders, sitting
each alone at the heart of his web, which to him is the entire world.[17]

Many stories can be told of Friday's tongue, but the true story is
buried within Friday, who is mute. The true story will not be heard till
by art we have found a means of giving voice to Friday.[18]

According to Susan, language is essentially human; hence, to be bereft of
speech is to dwell within an alienating silence that renders one less than
human, notwithstanding the fact that Friday is not also deaf, and therefore
cannot be likened to an animal existing in absolute isolation from others,
such as a whale or a spider (both species of which are capable of communi-
cating with other living beings, notwithstanding Susan's simile). Figuring
Friday as living in silence, Susan thus projects onto him *her experience* of his
absent voice, her experience of his silence.

In his analysis of Coetzee's *The Life and Times of Michael K*, Ato Quayson
suggests that the eponymous character's elective silence "might be taken as
an illustration of the autistic spectrum."[19] Although he briefly compares
the "scrupulous silence" that Michael and Friday both "enjoin upon themselves," the possibility that the latter's silence is volitional is nevertheless
discarded when Quayson asserts that Friday "is certainly mutilated and
without a tongue."[20] Whether Friday does or does not possess a tongue,
and whether he intentionally withholds speech from Susan or is physically
incapable of it, it seems crucial to counter her assumption that he inhabits
a self-enclosed world utterly cut off from language. Nowhere in *Foe* does
Friday "speak" to the reader in the manner of interior monologue. Yet the
sequence of interior spaces described in the novel's final section gives us to
believe that we might finally puncture the bubble that surrounds him. On
the brink of giving us his story only to take it away, *Foe* interrogates the
political platitude of giving voice, which is to say the devotion to speech
(whether literal or metaphorical) as evidence of the plenitude of political
recognition and presence, a view that rests on an uninterrogated conception of language as property.[21] As I argued in chapter 1, we do not own
language. Language is *leased* from language itself. Lessees without lessors,
we have no choice but to sign a "contract" that gives access to the language
in which we dwell but do not fully inhabit.

If language is originarily alien to each and every speaking human, even
to the vast majority of those whose tongues have not been violently severed,
then any claim of property or proximity that would permit language to

inhabit us, and vice versa, is undermined by an irremediable ex-habitation. Our mother "tongue" is inherently excised, cut off from ourselves. To speak of the loss of language in such general terms is undoubtedly to invite the *de rigueur* accusation of "eliding the specificity" of colonial violence, especially the physical brutality that results in the actual loss of one's tongue. Yet I wager that we must risk this allegation in order to demonstrate how the investment in language as property fails to recognize its own colonializing imperatives. As Derrida argues, "the master does not possess exclusively, and naturally, what he calls his language."[22] On the contrary, the master employs "force or cunning" through discursive, educational, and military means in order to reinforce the fiction of his linguistic possession.[23] He maintains:

> [We should not] efface the arrogant specificity or the traumatizing
> brutality of what is called modern colonial war in the "strictest defini-
> tion" of the expression. . . . But what if, while being attentive to the
> most rigorous distinctions and respecting the respect of the respect-
> able, we cannot and must not lose sight of this obscure common
> power, this colonial impulse which will have begun by insinuating
> itself into, overrunning without delay, what they call, by an expression
> worn enough to give up the ghost, "the relationship to the other"! or
> "openness to the other."[24]

Although Derrida is commonly interpreted as a thinker of difference and alterity, his emphasis here on the monolingual and the monological complicates this received view. Slogans such as "openness to the other" permeate scholarship on race, gender, class, sexuality, and postcoloniality. Many readers have often associated Derrida with a similar imperative to respect alterity, especially given the Levinasian echoes that resound throughout his work. Yet the alliterative, repetitive language of "respecting the respect of the respectable" clearly satirizes various discourses of alterity that piously call for a wholly nonviolent, nonappropriative relation to others.

In *The Monolingualism of the Other*, Derrida stresses the sovereignty that "tends, repressively and irrepressibly, to reduce language to the One," an "impassable, *indisputable* . . . inexhaustible solipsism," a drift toward the solitary self whose gravity he feels as strongly as does anyone else: "I am monolingual. My monolingualism dwells, and I call it my dwelling; it feels like one to me, and I remain in it and inhabit it. It inhabits me. The monolingualism in which I draw my very breath is, for me, my element."[25] That he freely admits his own complicity in this solipsism will seem anomalous to readers inclined to view him simply as an advocate of difference. Yet a preoccupation with an insuperable solipsism can be traced all the way back to *Voice and Phenomenon* (1967). When Derrida fleetingly remarks in *The Monolingualism of the Other* that our originary linguistic alienation introduces the phantasm of "hearing-oneself-speak in order to mean-to-say [*pour vouloir-dire*]," he is alluding to Husserl's discussion of interior monologue in the *Logical Investigations*, which is the central focus of *Voice and Phenomenon*.[26] For Husserl, interior monologue promises a realm of pure expression without communicative intent (when I "speak" to myself silently I do not indicate anything existent in the exterior world). Silent soliloquy involves "no function of indicating the existence of mental acts . . . for the acts in question are themselves experienced by us at that very moment."[27] Since communicative speech is both expressive and indicative for Husserl—since it requires the exteriorization of inner mental life in relation to an auditor who perceives such speech as indications or signs of the speaker's thoughts—silent speech has no need for indication because speaker and hearer are identical. Derrida's crucial contribution in *Voice and Phenomenon*, however, is to call into question the possibility of this space of pure interiority, utterly closed in on itself, a place of pure auto-affection: "A sign is never an event if event means an empirical singularity that is irreplaceable and irreversible. A sign that would take place only 'once' would not be a sign. A purely idiomatic sign would not be a sign."[28] Far from functioning as a sign beyond signs—which is to say a sign whose ownness restricts it to referring only to itself, to the *idios* or the *ipse* from which a sign would emerge and never leave itself—Friday's body *as* sign must lend itself to an ideality, or formal identity, that permits its repetition across innumerable empirical events and thereby exposes it to a temporal and spatial alterity that deprives it of any pure idiomaticity. In other words, a body could be a sign of something *other* than itself, as is the case with those gestures commonly referred to as "body language," but no body could be a sign simply and only for itself.[29]

Whereas Husserl believes that silent speech involves no hiatus between sign and meaning given that the latter is immediately present to the speaker as an intentional subject who says what he means because he means to say it, Derrida argues that "*a voice without différance, a voice without writing is at once absolutely alive and absolutely dead.*"[30] The voice of pure auto-affection whose meaning would be absolutely present to itself would dispense with writing (understood not merely as empirical inscription but as iterability in general). A silent utterance that could not be repeated—that was so idiomatic as to bear no relation to the past or to the future, an utterance absolutely tied to the singular life of the *ipse*—is absolutely dead. If the phantasm of auto-affection is produced in and through the reduction of an originary hetero-affection, then ironically the plenitude of a living speech that would remain absolutely close to me can only be achieved at the expense of denying language its life, its capacity to signify above and beyond my breath. This accounts for why the "truth" of Friday can only emerge in the form of a speech whose promise of transparent meaning opens onto an apocalypse that reveals nothing. The "words" that issue from the submerged Friday are literally breathless, as if to imply an utterly unique language that inspires no life beyond the instant of its emission.

In addition to the phenomenon of auto-affection, Derrida refers to another Husserlian concept in *The Monolingualism of the Other* when he asserts that, "far from sealing off anything, this solipsism conditions the address to the other, it gives its word, or rather it gives the possibility of giving its word."[31] Here the reference point is the principle of analogical appresentation. As I discussed in chapter 1, analogical appresentation names our elusive *experience* of others. This experience is not a matter of logical inference, but rather constitutes a spontaneous analogical transfer based on the perception of the other's similarity to myself. I experience others as inaccessible, but this does not render them absolutely unknowable. The quasi-monadic self is the condition of possibility for my indirect perception of *alter egos*.

Throughout his extensive body of work, Derrida frequently drew upon the twin Husserlian insights of auto-affection and analogical appresentation.[32] Yet, whereas the principle of appresentation survives more or less intact, auto-affection is submitted to a thorough dismantling by virtue of Derrida bringing the former to bear on the latter. Although Husserl maintains that interior monologue constitutes a closed sphere of pure expression in which the indicative function of language disappears, Derrida argues that this absence of indication is illusory: "[It only *appears* that] the subject does not have to pass outside of himself in order to be immediately affected by its activity of expression. My words are 'alive' because they *seem* [my emphasis] not to leave me, seem not to fall outside of me, outside of my breath, into a visible distance; they do not stop belonging to me."[33] So-called internal soliloquy occurs across time in relation to indicative signs whose capacity for repetition always bears the possibility that these signs will be severed from me. This severing is both spatial and temporal insofar as the signs that I employ in silent speech can be repeated to infinity (audibly or inaudibly) in other times and spaces. While repetition implies future externalization, a *possible* or *eventual* becoming nonproper and nonpresent, this a-proximity occurs anterior to my silent utterance given that it constitutes a "citational" act, a trace of prior linguistic utterances that do not have their origin in me.[34] Auto-affection is thus always already hetero-affection because my own self-relation is appresentive, that is, mediated through representational traces that precede and exceed me.

Such traces haunt *Foe*'s final chapter, which seems to hold out the possibility that the signifier can become "perfectly diaphanous by reason of the absolute proximity of the signified."[35] When Friday's mouth issues a stream of water rather than speech, this current washes over the unnamed narrator's face and then travels "northward and southward to the ends of the earth," as if having escaped its containment, the hitherto hidden interior of Friday now exposed to an infinitely expanding outside.[36] Yet we can no more access the truth of his interior life than we can anyone else's. The novel thus stages this incomplete, appresentational contact when it welcomes us into his home, only to be immediately borne away by the stream emanating from his mouth. The home of Friday thus designates an impossible place where absolute interiority coincides with absolute exteriority. An "inside" that appears entirely enclosed on itself, an inside without any outside, paradoxically equates to an infinity without borders. When Friday finally "speaks" to us, he does so from the oceanic depths where "words" flow outward toward infinity, toward an ostensibly boundless outside that nevertheless envelops itself by virtue of having no borders. "Bodies are their own signs" only in this breathless place of pure interiority without verbal or written language.

Coetzee's depiction of a body that signs in and for itself thus implies a pure materiality absolutely liberated from the representational domains of speech and writing for which Susan has doggedly sought Friday's inclusion, as if Friday finally responds to Susan's demand for answering speech by mouthing, "I have no need for speech or writing, thank you very much." In an often-cited interview with David Attwell, Coetzee remarks: "Friday is mute, but Friday does not disappear, because Friday is body. If I look back over my fiction, I see a simple (simple-minded?) standard erected. That standard is the body. Whatever else, the body is not 'that which is not,' and the proof that it *is* is the pain that it feels. The body with its pain becomes a counter to the endless trials of doubt (One can get away with such crudeness in fiction; one can't in philosophy, I'm sure)."[37] For Coetzee, the conclusion of his novel attempts to achieve closure by putting a stop to ceaseless skepticism: "Is representation to be so robbed of power by the endlessly skeptical processes of textualization that those represented in/by the text—the feminine subject, the colonial subject—are to have no power either?"[38] Too much skepticism is said to weaken representation on the one hand, yet nevertheless invest it with an ironic power to deprive Susan and Friday of power on the other. Coetzee goes on to suggest that the novel's "preemptory ending . . . close[s] the text by force" rather than accept "the prospect of endlessness."[39] Taken together, the question of oppositional power and that of the mandated conclusion imply that the ending aims to force open an absolute extra-textual outside (or absolute inside, which amounts to the same thing) that would allow the previously muted power of the female and colonial subjects to speak. Yet if the restoration of the other's power is predicated on ceasing with textuality altogether—as if Friday could only speak once Coetzee stops writing—then this absence of representation would amount precisely to the annulment of any oppositional power. What force could an unrepresented or unrepresentable power wield?

Whereas Benita Parry reproaches Coetzee for allegedly straying from his critique of political oppression by promoting a "non-linguistic intuitive consciousness," an "ineffable" language that credits Friday with "mystical properties and prestige," Coetzee's response to Attwell attempts to situate the ending of the novel squarely with the framework of postcolonial critique by lifting the hitherto suspended access to Friday's body.[40] Is Coetzee guilty of the crudeness or simple-mindedness that worries him by endlessly deferring the presence of Friday's voice only to supplant it with the body? Or should we avoid the temptation to exploit his self-deprecating comments and ask instead whether a novel otherwise so keenly attuned to the interplay of multiple linguistic stratums ultimately signifies, in a strictly serious, wholly nonironic fashion, a corporeality utterly removed from speech and writing? Suffice it to say that Friday's body is necessarily inscribed within a text that produces the fiction of speaking through the voice of a disembodied narrator, which means that the home of Friday can only *represent* a realm ulterior to inscription, an impossible territory of nonlanguage that nevertheless takes place *within* language.

This pure interiority toward which the conclusion of *Foe* gestures, a sphere that dispenses with linguistic signs, thus bears a striking resemblance to what Derrida characterizes as the "prior-to-the-first time of pre-originary language," the invention of an absolute idiom, a pure monolanguage that would thoroughly align with the self, that would require no detour through alterity.[41] Of course, this utterly solipsistic language does not and cannot exist. A language without alterity is no language at all. Yet it survives as the memory of what never was: the absolute possession of "my" language. That we do not possess language so much as we are possessed by it means that its spectrality constitutes the "degree zero-minus-one of memory."[42] One does not start with *one* native language onto which are grafted additional foreign tongues. One starts with zero, a buttonhole of sorts, an *abiding* absence and dispossession that the acquisition of a native language cannot fill. The zero-minus-one of this language before language names the empty placeholder of a purely idiomatic language where bodies would have no need of signs.

Wanting to Say

Just as interior monologue's would-be transparency of meaning is undone by the mundanity of indicative signs against which no soliloquy can isolate itself, the home of Friday names precisely an unhomely (*unheimlich*) space submerged in a cloud of murky water that occludes access to Friday's "true story." The question that the unidentified narrator asks Friday—"what is this ship?"—echoes Foe's speculation that the ship was a slave vessel not a merchantman.[43] Friday's lack of response, however, means that we can neither confirm nor deny Foe's conjecture.[44] In addition to withholding the ship's identity (merchant or slave vessel), the novel also does not resolve the question of whether Friday's tongue is indeed absent, and if so, who removed it. Susan comes to distrust Cruso's account that slavers committed the mutilation, and tries unsuccessfully to extract the truth from Friday in order to confirm her suspicion that Cruso himself executed the cruel act. If Susan doubts Cruso regarding the perpetrator of Friday's maiming, however, she nevertheless takes him at his word that Friday has been deprived of the word due to his mutilation. Yet on what basis ought Susan to accept that Friday has no tongue? Given that she refuses to examine his mouth, his verbal silence—save for saying "ha-ha-ha" when Cruso commands him to say "la-la-la"—provides the only evidence to support her conviction.[45] Susan speculates wildly as to the cause of Friday's absent tongue. While she steadily maintains that the culprit is either Cruso or a slave trader, she briefly ponders whether his tongue was removed in infancy, "at the age when boy-children among the Jews are cut. . . . Who was to say there do not exist entire tribes in Africa among whom the men are mute and speech is reserved to women? Why should it not be so? The world is more various than we ever give it credit for."[46] This analogy with circumcision morphs into a parallel with castration later in the novel when Susan likens Friday to a "gelding" and she wonders whether Cruso may have spoken about Friday's absent tongue metaphorically to indicate "a more atrocious mutilation . . . a slave unmanned."[47] However, as with the question of whether his tongue has indeed been excised, the novel does not provide definitive proof that Friday has been castrated.

Despite the curiosity that Susan's imaginative ruminations display, they nonetheless presume the presence of some absence, as it were, a lack whose location and cause may be uncertain but whose existence is undeniable. Susan takes it as a given that Friday lacks either a tongue or a penis or both. Moreover, her revulsion toward his ostensible mutilation is not solely an expression of moral outrage against the cruelties of slavery. For even when she speculates that the amputation might belong to an African custom that deprives men of speech—or later when, making little headway in teaching Friday to write, she muses that perhaps Friday is silently laughing at and mocking her "efforts to bring him nearer to a state of speech"—Susan never questions the value of speech over silence, nor does she consider that the latter does not equate with the absence of the former.[48] That she suspects Friday may be mocking her is the closest she comes to ascribing him any agency, as if to imply that his silence may be volitional. When Susan fails to engage Friday in a musical duet that she had hoped would have supplemented his verbal deficiency, she wonders if "it might not be mere dullness that kept him shut up in himself, nor the accident of the loss of his tongue, nor even an incapacity to distinguish speech from babbling, but a disdain for intercourse with me."[49] Her desire for "answering speech," even if it must finally take the form of writing, follows the circuit of auto-affection: a desire to hear herself speak. To account for this ironic reversal whereby the desire for the other's speech amounts to a desire to hear oneself speak, we must grasp how Susan conceives of language as a gift to bestow upon the other, a conception that discloses a possessive investment in language notwithstanding its irremediable dispossession. The specter of Friday's absent tongue thus provides a useful surrogate for her own linguistic buttonhole. Appointing herself as both guardian and teacher of Friday, she adopts the role assumed by Defoe's Robinson Crusoe, who expresses a desire to "*make* [my emphasis]" Friday "speak, and understand me," a desire all but abandoned by Coetzee's Cruso, who has little interest in teaching his companion to understand more than a few simple phrases.[50] Puzzled by Cruso's indifference to intersubjective communication, Susan explains to Foe that "life on the island, before my coming, would have been less tedious had he [Cruso] taught Friday to understand his meanings, and devised ways by which Friday could express his own meanings, as for example by gesturing with his hands or by setting out pebbles in shapes standing for words."[51] Who would dare object to the apparently incontrovertible common sense of this affirmative valuation of language? Who would want to champion the almost autistic quality of Cruso's desire for silence and insularity? Yet perhaps we need not choose between Susan's naive quest for intersubjective speech as self-presencing truth or Cruso's solipsistic self-encirclement. If auto-affection is always already riven by hetero-affection, if appresentational (spatial) and representational (temporal) traces expose my monadic self to an exteriority that mediates both my self-relation and my relation to others, then neither pure self-containment nor pure openness toward alterity is possible.

When Susan says that she wants to give voice to Friday, she means that she wants him to want to speak. As Derrida underscores, desire is not merely accidental or exterior to signification; rather, meaning and desire are indissoluble. To stress this connection, he translates the German term *bedeutung* (meaning) into French as *vouloir-dire*, literally "wanting-to-say." What Derrida wants to say is that meaning is always bound up with volition. This will and drive for stability and transparency is manifest particularly in Husserl's insistence on meaning as intentional, which he argues excludes facial expressions and other unconscious, bodily gestures from the sphere of signification. Susan's volition thus warrants translation into the following (admittedly nonidiomatic) English: she wants to say that she wants Friday to want to say. Yet how are we to separate her wanting to say from his wanting to say? How does Susan even know that Friday wants to say anything? As Wendy Brown remarks: "[I]f the silences in discourses of domination are a site for insurrectionary noise, if they are the corridors we must fill with explosive counter-tales, it is also possible to make a fetish of breaking silence. Even more than a fetish, it is possible that this ostensible tool of emancipation carries its own techniques of subjugation—that it converges with non-emancipatory tendencies in contemporary culture."[52] Brown is particularly concerned with a certain "pre-Foucauldian" tendency within feminist politics to view speech as either expressive or repressive. On this view, speech either expresses freedom and selfhood or leads to further oppression through hate speech, pornography, or harassment: women's "truth" or men's "truth." Susan clearly employs an expressive/ repressive model of speech, though the gendered opposition between female expressivity and male repressivity is inverted and reframed in racialized terms. As a white woman eliciting speech from a black man, Susan is not unaware of the power dynamic that subtends their relationship: "I tell myself that I talk to Friday to educate him out of darkness and silence. But is that the truth? There are times when benevolence deserts me and I use words only as the shortest way to subject him to my will. At such times I understand why Cruso preferred not to disturb his muteness. I understand, that is to say, why a man will choose to be a slaveowner. Do you think less of me for this confession?"[53] Confessing to Foe her desire to dominate Friday, Susan ironically inserts herself into the regulatory discourse of confessional truth by assuming the subjugated role of the confessee. In fact, this passage is doubly ironic: she expresses an awareness of the potentially coercive function behind her demand for Friday's expression, yet she betrays a correlative compliance to discursive power by openly confessing her desire to a white man.

Whereas Susan conceives speech as divided between expressive and repressive forms of power, she views silence as unfailingly repressive. Her fleeting cognizance of discursive and confessional power thus fails to dislodge the equation of silence with oppression to which she largely subscribes. However, as Foucault argues in *The History of Sexuality*:

> Discourses are not once and for all subservient to power or raised up against it, any more than silences are. We must make allowance for the complex and unstable process whereby discourse can be both an instrument and an effect of power, but also a hindrance, a stumbling-block, a point of resistance and a starting point for an opposing strategy. Discourse transmits and produces power; it reinforces it, but also undermines and exposes it, renders it fragile and makes it possible to thwart it. In like manner, silence and secrecy are a shelter for power, anchoring its prohibitions; but they also loosen its holds and provide for relatively obscure areas of tolerance.[54]

In her reading of this passage, Brown clarifies that silence may resist regulatory discourses, but "practices of silence are hardly unfettered."[55] In this regard, we should avoid the naive temptation to view Friday's silence as occupying an apolitical space utterly insulated from colonial power. Not only does silence not equate to the absence of speech—as the activity of internal monologue demonstrates—but this ostensibly pure expressivity cannot separate itself from the indicative world, which is to say that it cannot withdraw from alterity *tout court*. Although Brown identifies as *potentially* fetishistic the compulsion to speak in the name of an emancipatory politics, this characterization is less a nod to Freud than it is a cautionary warning to eschew excessive devotion to the promise of liberatory speech. Yet, the wanting-to-say of signifying practices is *inherently* fetishistic insofar as the desire to align saying and meaning—whether the meaning of internal soliloquy or of spoken utterance (my own or that of the other)—always strives for an unachievable plenitude.

The Dumb Sovereign

The resemblance that Susan identifies between her demand for responsive speech and the master's subjugation of the slave recalls Charles Chesnutt's short story, "The Dumb Witness" (1897).[56] Set in the nineteenth-century American South, the story centers on a slave named Viney whose tongue is maimed by a master named Murchison in retaliation for sharing with his betrothed a secret that causes her to call off the wedding. The content of the secret is never revealed, though the story strongly indicates that Murchison and Viney are involved in a miscegenous, perhaps even incestuous, relationship. She is described as a "young quadroon" with a "dash of Indian blood," and is also said to be "of *our* blood [my emphasis]," a phrase that ambiguously signifies both race and family, implying that she is genetically kin in addition to being part white.[57] Murchison manages the property for an absentee uncle who wills it to his nephew shortly before the former's death. The location of the will is known only to Viney, but her mutilated condition, combined with her illiteracy and Murchison's failed efforts to teach her to write, prevents her from disclosing its whereabouts.

The story begins in the postbellum South when a white enterprising Northerner named John calls on the old Murchison property to inquire about purchasing some lumber. John observes Murchison entreating Viney to reveal the location of the will, only to receive a response of "discordant jargon."[58] We learn that this scene has played out for many years to no avail and Murchison eventually dies without receiving his inheritance. The following summer when John pays Murchison's son a visit, he is greeted by Viney, who miraculously seems to have recovered the power of speech. As John's coachman Julius reveals, however, she never lost her capacity to speak, but had feigned its loss in order not to reveal the location of the will, which turns out to have been hidden in the seat of a large oak chair where Murchison sat for many years.

What initially appears to depict solely a horrific scene of violence and linguistic deprivation turns out to have also been a contest of wills whereby Viney chooses silence. Beyond the obvious parallels with Coetzee's novel, "The Dumb Witness" shares with *Foe* a preoccupation with interiority and unfathomability: namely, the numerous holes that Murchison burrows in the ground in hollow pursuit of his will, holes that metonymically evoke the mouth to which Viney woefully points when he asks her to reveal the will's location. When Murchison is not "digging, digging furiously" in the ground, like a dog struggling to unearth a bone, he barks orders at Viney to reveal the location of his "princely inheritance."[59] That he never looks under the large oak throne from which his demands vainly issue raises the question: who is the real dumb witness of the story?[60] Is it not the witless master who fails to ascertain that he is sitting on his own will? This power reversal speaks to the unexpected symmetry that Derrida captures with the phrase *la bête et le souverain*: the beast and the sovereign.[61] Derrida notes that the homophony of *et* and *est* permits us to hear this phrase as the beast *is* the sovereign. The phrase thus bears witness to a mute distinction that can only be "heard" textually. This linguistic coincidence reflects a larger political structure by virtue of which the beast and the sovereign occupy parallel positions outside the law: the sovereign is "above" the law insofar as he exempts himself from it, the animal is said to be "below" the law, ignorant of it and therefore deprived of freedom and agency. Viney is similarly below the law because slaves are not considered legal subjects. Yet she no doubt derives satisfaction from knowing that the sovereign physically sits above the law—or at least above the documents that certify his legal entitlement—only insofar as he rests ignorant of it.

The story's tragic-comic irony derives not only from the revelation of the will's proximity to the master, but also from our retrospective realization that his repeated pleas to Viney betray fragility rather than strength. Viney weakens the master precisely by enjoining *him* to speak, to engage in a "dialogue" that leaves him increasingly exasperated. As discussed in chapter 1, the "reason of the strongest" exempts one from the obligation to provide explanations. Sovereignty means never having to give reasons. As Derrida observes, pure sovereignty "always keeps quiet in the very ipseity of the moment proper to it, a moment that is but the stigmatic point of an indivisible instant."[62] Sovereignty undermines itself as soon as it speaks, as soon it announces itself as such. That sovereignty is absolute as long as it is "dumb" is also to say that it is never absolute. Viney weakens the auto-affective fantasy of the master by forcing him to beg for assistance. To be sure, the would-be sovereign *almost* always holds the upper hand within racist power structures even if he is *always* almost sovereign, even if the purity of his sovereignty is compromised from the beginning.[63] By contrast, the exceedingly tenuous character of slave sovereignty is registered when Viney chooses to remain on the plantation after slavery is abolished, as well as when she discloses the will's location to Murchison's son, thereby ensuring the transfer of inherited white wealth.

The role that Julius plays in the story is equally relevant to the nexus of sovereignty and silence. Chesnutt's stories typically employ a framing device in which John temporarily cedes the standard English of his narrative voice to Julius, who relates in black dialect various tales of antebellum slavery. Eric Sundquist suggests that Chesnutt's anomalous decision to withhold Julius's voice stresses "American culture's exclusion of the folk-loric oral world of black culture" as well as implies an analogy between the different modes of silencing to which Julius and Viney are subjected.[64] The dialect that Viney employs when she finally speaks bears out this connection. When John asks Viney if the young Murchison is home, she replies: "Yas, suh . . . I'll call 'im."[65] As Richard Broadhead notes, an earlier version rendered her speech in standard English. Speaking in dialect, Viney resurrects the black vernacular that John has suppressed by claiming to retell the tale in "orderly sequence," as if to associate Julius's dialect with Viney's "meaningless cacophony."[66] Sundquist suggests that Julius "has played 'dumb' all along," but the intentionality of this play remains ambiguous.[67] Julius is not "playing" in the same way that Viney feigns an inability to speak. As Sundquist makes clear, Julius's silence is imposed through John's act of cultural appropriation. Yet Julius may indeed be intentionally playing dumb in a manner that Sundquist does not consider. John informs us early on that Julius was "ignorant" of "some of the facts" of the story, but if the former gleaned these facts from "other sources," then the information they provided must not relate to Viney's linguistic capacity, which Julius divulges to John when he returns to the plantation.[68] Does Julius learn the truth only after Murchison dies, or was the former in on the ruse of Viney's dumb show all along, in which case John turns out to have unwittingly played the fool? Julius has held his tongue regarding Viney's true condition, the knowledge of which we might surmise passed freely among the black community through what Booker T. Washington called the "grape-vine telegraph," a word-of-mouth network hidden from whites.[69] According to Washington, during the Civil War "often the slaves got knowledge of the results of great battles before the white people received it."[70] Viney's name communicates not only this tangled network of speech withheld from whites, but also Chesnutt's "The Goophered Grapevine," a story in which Julius fails to dissuade John from purchasing a vineyard whose grapes had sustained the former's income for many years. Julius attempts to protect his interests by telling John that the grapes have been "goophered" (bewitched) by a conjure woman. As Robert Bone suggests, "Julius is a kind of conjurer, who works his roots and plies his magic through the art of storytelling."[71] Viney can also be read as a "conjurer" who employs silence rather than speech to challenge the authority of the master. At bottom, she thrusts him off his "ancestral seat," leaving him only to dig his own hole(s).[72] Chesnutt's editor, Walter Hines Page, reportedly excluded "The Dumb Witness" from the first edition of the collection of stories published as *The Conjure Woman*. Claiming that it lacked both the supernatural element of conjure and the black vernacular storyteller, Page apparently was unable to hear the silent conjure that winds its way through Viney's doleful yarn.[73]

"There is no world, there are only islands"

Notwithstanding their disproportionate power, the silence of the sovereign and the silence of the "beast" disclose a shared impurity, an impossible absolution thanks to which silence is never utterly silent, is never simply the opposite of speech. The plenitude of an unfettered speech or silence would only be achievable in the imaginary space of an absolute solipsism, the preoriginary language of an unqualified private idiom, the sovereign sphere of a nonshareable, indivisible language. In *Foe* Cruso's apathy toward being rescued, as well as his general disinterest in verbal communication, can perhaps be too easily scorned for its apparently solipsistic or autistic traits. As Susan remarks, "I used once to think, when I saw Cruso in this evening posture, that, like me, he was searching the horizon for a sail. But I was mistaken."[74] Later Susan recalls his reflection that "the world is full of islands," a remark that parallels Derrida's claim in *The Beast and the Sovereign*, volume 2:

> [No one] inhabit[s] the same world, however close and similar these living individuals may be. . . . between my world and any other world there is first the space and the time of an infinite difference, an interruption that is incommensurable with all attempts to make a passage, a bridge, an isthmus, all attempts at communication, trope, and transfer that the desire for a world or the want of a world, the being wanting a world will try to pose, impose, propose, stabilize. There is no world, there are only islands.[75]

These comments may seem to advance a purely isolationist view of the world, yet they can be interpreted thus only by reading them in isolation. If we bridge these comments to the sentence that immediately precedes it, however, we confront the following paradox: "Incontestably, animals and humans inhabit the same world. . . . Incontestably, animals and humans do not inhabit the same world."[76] What Derrida says of interspecies relations also applies to intraspecies ones. All humans both do and do not inhabit the same world. Since I never have direct access to you, to your world, I can only ever carry you within "my" world. I can only bear the trace of you and your world within the worldless world in which I dwell but to which I only have partial access, this veiled world whose opacity owes precisely to the existence of other worlds, spheres whose presence I encounter only appresentationally through the traces those worlds imprint on me.

"There is no world, there are only islands" thus amounts to saying: there is no one world that we all share, a world without *différance*, a world without the interruptions of time, language, and iterability, however much the flow of language may appear to connect us from one end of the earth to the other. The phantasm of one united world to which all subjects belong promotes the worst kind of (quasi)solipsism, a pregiven world that fully transcends those living subjects who are thereby reduced to a second-ary role of participation without intentionality, the latter understood in the phenomenological sense as consciousness directed toward an alterity that we experience as *for us*, as in some sense constituted by and dependent on us. The principle of one shared world is solipsism by stealth: it promises to transcend the apparent enislement of the monadic ego through a lan-guage of unity that disavows our appresentational relation to alterity. To say that "we all inhabit the same world" is really to say, "we all inhabit *my* world." However I imagine it, the world that I posit as *the* world in which I dwell with others masks the monospheric character of this world—the extent to which this world originates in me.

This immanent transcendency that Husserl identifies in our relations with others ironically unsettles his conception of interior monologue, which presumes precisely a subject who is proximate and present, here and now, to its own meanings. That this view of signification is fundamen-tally fetishistic does not mean that we can or should abandon the desire to mean to say. The absolute coincidence of sign and meaning may be an impossibility, yet the abandonment of all signifying practices is neither desirable nor achievable. Nevertheless, the desire for the absolute diapha-neity of the sign can and ought to be curtailed precisely in the name of the alterity that no politics or ethics can do without—in the name, that is, of a *weaker solipsism*.

When Husserl writes that the mental acts of interior monologue are "experienced by us at that very moment," his own language undoes its aspiration to pure immanence. Husserl employs the German phrase *im selben Augenblick*, "in the blink of an eye," an expression that Derrida seizes upon in order to counter the assertion of an absolute idiom that could bracket itself both temporally and spatially. No less so than in spoken lan-guage, interior monologue requires that I "speak" across temporal moments that introduce a hiatus, no matter how minimal or immeasurable, between speaking and hearing. "In the blink of an eye," my words become available to iterability and therefore stop being mine, even though, of course, they were never fully mine to begin with. While Husserl maintains that every punctual now bears within it a retentional element of the past it has just superseded and a protentional (anticipatory) element of an incipient future, he wants to distinguish retention from memory by insisting on the former's nonrepresentational status. The retentional phase would open onto an immediate relation to the past that is subsequently represented through secondary memory. Yet Derrida remains skeptical that the retentional phase can be isolated from representation: "As soon as we admit this continuity of the now and the non-now . . . we welcome the other into the self-identity of the *Augenblick*, non-presence and non-evidentness into the *blink of an eye of the instant*. There is a duration to the blink of an eye and the duration closes the eye."[77] As a visual trope, "the blink of an eye" stresses the spatial, appresentational dimensions of this alterity, whereas the temporal interval between the opening and closing of the eye introduces the possibility of repetition in the wake of my absence.

Susan employs a trope identical to Husserl's when she gives in to Cruso's unwanted sexual advances, even though she believes she could have overpowered him with her superior physical strength. After their encounter she wonders if she should regret what she has allowed to occur before quickly settling on acquiescence: "We yield to a stranger's embrace or give ourselves to the waves; for the blink of an eyelid our vigilance relaxes; we are asleep; and when we awake, we have lost the direction of our lives. What are these blinks of an eyelid, against which the only defence is an eternal and inhuman wakefulness? Might they not be the cracks and chinks through which another voice, other voices, speak in our lives? By what right do we close our ears to them? The questions echoed in my head without any answer."[78] Susan presents hospitality as an ethical obligation whose passivity is veritably Levinasian in its apparent selflessness. Given the context of male sexual domination in which she voices this passivity, her selflessness is especially troubling. She frames her refusal to resist male aggression as a form of hospitality, as if resistance to Cruso would have been unethical. In this sense, her thoughts are prescriptive. Yet she also hints at the impracticability of an "eternal and inhuman wakefulness." Aligned with the sexual act, this relaxed vigilance implies the possibility of additional *openings*, bodily orifices through which the other physically enters, thus supplementing the sites of penetration that I enumerated earlier (eye sockets, mouths, ears, rooms, and deep-sea shipwrecks). Is it not so much a question of whether we have the *right* to close our ears to others as it is a question of whether closing ourselves off to the other is finally possible?

Certainly a woman not only has the right to make her herself unavailable sexually, but is also physically able to do so—at least under "normal," noncoercive or nonviolent circumstances. Yet unlike sites of sexual penetration, our ears are completely open and exposed, a vulnerability that Husserl's optical metaphor disavows. We can always close our eyes, but we can at best dull or mute the sounds that forever penetrate this most passive of our senses. In contrast to Husserl, Coetzee's synesthetic trope introduces precisely the hiatus that Derrida identifies: a lapsed awareness that allows the voices of others to be seen or heard. Ironically, Susan ruminates about such external voices through interior monologue: "The questions echoed in my head without any answer." She hears herself speak, pondering whether one has the right to shut out other voices entirely. Yet if a constant wakefulness is not humanly possible to maintain, then we are fissured by alterity from the beginning. Prior to the question of our *right* to inhospitality, a certain originary "yes" to alterity conditions the possibility of our deciding whether we are ethically authorized to mute our ears to others.[79] We must have already "said" yes to others in order to be able to say no. The *o* that seems to circumscribe us is always punctured by the traces of others whom we can never entirely envelop within us. We thus do not live according to Susan's vision of the whales, floating leagues apart from one another. We live more like spiders, sitting at the center of our webs, the 0° position from where we fall prey to other webs, worlds, or islands, an unhomely home insofar as we are drawn out from ourselves, exposed—shipwrecked, for better or worse, on one another's shores.

CHAPTER 3

The Gravity of Melancholia
A Critique of Speculative Realism

I have neither up nor down, like the squirrel climbing up
and down horizontally, the form of my world, a literature that is
apparently, like the very look of my writing, cosmonautical,
floating in weightlessness.

—JACQUES DERRIDA, "Circumfession"[1]

Gravity and Melancholia: two forces of attraction that weigh on us, the former by tethering us to the ground, the latter by gripping us in perpetual, unfinished mourning. Gravity as a physical force is a necessary condition *of* us. Our survival depends on it. Melancholia as psychological depression, by contrast, is a condition *for* us. It radiates outward from the center of an ego impoverished by grief and loss. We might well long to escape this attraction that propels us into a fixed orbit around the love objects that we have lost, but who could say the same for the physical laws of the universe? Gravity is a force of nature that we not only cannot escape, but that physically and psychically grounds us, assuring us (no matter how falsely) of our centeredness and immovability.

In Lars von Trier's *Melancholia* (2011) and Alfonso Cuarón's *Gravity* (2013), however, the twin forces of gravity and melancholia act unpredictably.[2] Melancholia also names a planet that falls out of its orbit and collides with Earth. As it approaches, Melancholia weakens the gravitational pull of our small planet, producing a weightlessness that is not altogether unwelcome: Justine (Kirsten Dunst)—whose crippling depression weighs her down to the point that others must help her out of bed and assist her in bathing, who characterizes her condition as "trudging through this gray woolly yarn"—is notably lifted out of her melancholia the closer the rogue planet careens toward Earth. In *Gravity*, astronaut Ryan Stone (Sandra Bullock) experiences a physical weightlessness that becomes utterly terrifying when the space shuttle she is attempting to repair is struck by satellite debris. Shortly after the collision, Ryan reveals her own melancholic condition when she confides in the only other survivor of the shuttle's destruction, Matt Kowalski (George Clooney), that she lost her daughter Sarah in a sudden, accidental death when she fell down and hit her head. Perhaps not by accident would a mother whose daughter's death can be attributed to gravity itself seek to diminish its force, if only to find her own life ironically threatened by the forces of gravity when she finds herself in the midst of a cosmic collision.

What does it mean that melancholia can attract such diametrically opposed forces as gravity and weightlessness? What is the gravity of melancholia that it can feel at once so heavy and so buoyant? Inversely, what is the melancholia of gravity? The grief that Ryan suffers is no less encumbering than the "woolly yarn" of Justine's melancholia, notwithstanding the weightlessness in which the former has sought refuge. To experience loss is to feel both weighed down and strangely detached from one's environment. Far from being opposed, buoyancy and ponderance express two sides of the same melancholic condition.

In Freudian terms, both mourning and melancholia are characterized by ambivalence, by a simultaneous attraction and repulsion in relation to the lost object. Freud maintains that one must release oneself from the orbit of the lost love object in order to complete the grieving process: "When the work of mourning is completed the ego becomes free and uninhibited."[3] The end of mourning occurs, according to him, when the ego displaces the love for the lost object onto a new one. In this sense, Freudian theory endorses a certain zero gravity in relation to lost objects, coupled with a regalvanized attraction to new ones.

The question of object attachment is now a central theoretical concern due to the emergence of speculative realism, which endorses what Graham Harman describes as an utterly "non-relational conception of the reality of things."[4] Since Immanuel Kant, continental philosophers have generally held that our knowledge of the world is never direct and unmediated. We can only ever have access to phenomena (representations) of noumenal reality (things-in-themselves). Quentin Meillassoux dubs this view "correlationism" in order to critique its apparent privileging of the human-world relation. Notwithstanding Kant's claim that the counterintuitive thinking that led to Copernicus's discovery of Earth's orbit is analogous to the counterintuitive notion of our mediated relation to reality, Meillassoux charges Kant with unleashing a "Ptolemaic counterrevolution," reinscribing the human as the center of knowledge.[5] Inspired by this critique of correlationism, a group of theorists including Harman, Levi Bryant, Ian Bogost, and Timothy Morton have promoted a new philosophical approach called object-oriented ontology (OOO) that seeks to fully renounce human privilege altogether in order to place all beings (human, animal, plant, and thing) on the same ontological footing. OOO maintains that we can posit a world that exists entirely independent of us in which things are granted their true "ontological dignity."[6]

Gravity and *Melancholia* both direct our attention toward the nonhuman things that inhabit our world. Although cinema always depicts things, these two films tend to horizontalize the relation between human and thing, as well as call into question the privileged position from which the human judges the relative significance of things. In the wake of *Melancholia's* Wagnerian overture, for instance, we witness a certain leveling of the distinction between large and small objects when, immediately after depicting Earth's destruction, the scene shifts to a narrow country road where an enormous limousine is attempting to navigate an extremely tight curve. Inside are the newlyweds, Justine and Michael (Alexander Skarsgård), en route to their wedding reception at her sister's remote estate. While they each take turns in the driver's seat after the chauffeur proves incapable of rounding the corner, we observe the car lurch backward and forward as they both attempt to guide the unwieldy vehicle. Ultimately Justine steers the car too far toward the road's edge and the car's bumper hits a stone boundary marker. In addition to commenting on the overly lavish rituals of modern weddings, this scene derives its farcical value from the contrast it stages between interplanetary collision and the apparently trivial impact of a limo with a small rock. The limo and the stone thus rehearse in mundane terms the previously observed collision of the immense Melancholia with our tiny Earth, an impact that we as spectators witness from an extraterrestrial perspective, that is, from the standpoint of would-be but impossible survivors. The film's solemn opening yields to subtle mockery, as if to imply either that Melancholia's collision with Earth is no more significant than the limo's impact with the stone, or that the latter collision is as catastrophic as the former.

We witness this force of things in another scene in *Melancholia* when Claire is informed that none of the wedding guests correctly estimated the number of beans in a jar. Dismissing the game as "incredibly trivial," Claire shocks the bean-counting wedding planner by refusing to give any weight to these tiny materialities. OOO may insist that these objects are just as consequential as any other, but Claire not surprisingly displays more concern with the threat posed by Melancholia later in the film. As the planet looms large on the horizon, Claire carries Leo across a golf course and suddenly endures a deluge of hail. Confirming that things act in a manner that encumbers human agency, the hail causes her to falter and set him down. The hail looks suspiciously like the white beans whose ontologies she earlier dismissed, as if they are wreaking revenge on human narcissism. Indeed, their agency puts humans and objects on the same footing precisely by causing Claire to lose her footing on a hill of "beans." Whether tiny or big, nonhuman objects repeatedly display agency in *Melancholia* in a manner that thwarts the humanist disregard for nonhuman action.

Gravity is perhaps even more overtly object oriented, given both the sheer variety of different entities with which Ryan collides and the stereoscopic representation that lends them their manifest thinginess. The appeal of the film partially lies in stereoscopy's most familiar and illusory effect: that of objects projecting into the foreground, beyond the flat plane of the screen, as if beckoning the viewer to reach out and touch these things-in-themselves. Ryan, on the other hand, exhibits a persistent nonrelationality despite her briefly coming into contact with various pieces of space debris. When she and Matt attempt to attach to the Russian space station, for example, her foot is caught in the lines of a Soyuz parachute. While she manages to grab hold of the tether to which Matt remains attached, his weight pulls him away from the space station. He then unclips himself from the rope and floats away so that she can survive. "Ryan," he says, "you're going to have to learn to let go." Weightlessness provides no more protection from death than does gravity. In contrast to the abrupt collision of Sarah's tiny mass with Earth's much larger one, Matt will drift toward his death perhaps without ever coming to rest. His instruction "to let go" clearly implies a larger moral allegory; she must learn to accept not only his loss but also that of her daughter. This attachment is complicated by its apparent obverse. Struggling to correlate with others, both human and nonhuman, Ryan is melancholically unable to connect with any object.

In another scene Ryan and Matt return to the space shuttle to search for survivors. We see a number of objects emerging from the damaged fuselage, including a baseball hat, Rubik's Cube, dental retainer, headphones, screwdriver, and a figurine of the Looney Tunes character Marvin the Martian (Figure 2). Similar to the red shoe whose loss Ryan's daughter deeply lamented, but which her mother located after her death, these mate-

Figure 2. Objects from the space shuttle in *Gravity*.

rial traces are all that remain of the astronauts. As with the bolt that Ryan drops when she is repairing the satellite or the tears that she sheds once she discovers that the Soyuz is depleted of fuel and all hope of returning to Earth seems lost—tears that Cuarón's script likens to "tiny satellites orbiting her face"—these objects of ordinary human life float toward negative parallax before drifting off beyond the screen's borders.[7] To feel the effects of negative parallax as a spectator of *Gravity* is to share Ryan's experience of orienting toward objects that repeatedly slip away, like so many inaccessible Kantian noumena. The figurine of Marvin the Martian in particular betrays this tension between appropriation and alienation. Advancing near us before receding toward the right side of the screen, this figure of alien life presumably stimulates the childhood memories of many spectators who might find solace in its familiarity or humor in its ironic appearance, if only to experience an affective shift when, literally out of left field, the upturned face of a dead female astronaut suddenly emerges. Her frozen visage collides with the equally solid glass of Ryan's helmet, as if presaging a becoming thingly-in-death that may befall Ryan, too.

The objects that OOO inventories, by comparison, do not so much represent lost human life as they do an injunction to lose the human-centered world, thereby disavowing the loss of the thing-in-itself bequeathed to us by Kantian "correlationism." OOO's "flat ontology," as Bryant calls it, insists on giving equal ontological weight to all objects precisely on the grounds of a certain groundlessness, on the basis of a zero gravity whereby no object bears any greater mass, any more attractive pull, than any other.[8]

Ryan's analogous buoyancy is coextensive with her professed habit of driving aimlessly around her home town of Lake Zurich, Illinois. As she confides to Matt, she received the news of her daughter's death when she was driving her car: "Ever since that's what I do. I wake up, I go to work and then I just drive." Driving allows Ryan both to reenact the trauma and to imagine through endless repetition the possibility of a different outcome, one in which her movement would not have been arrested by the news of her daughter's death. Inertia and weightlessness cooperate: driving around town or orbiting around Earth, Ryan seeks the twofold suspensions of unrestricted momentum and zero gravity.

The storyline stresses the parallels between driving and flying when, at the end of the film, Ryan leaves the Soyuz for the Chinese space station (in whose escape pod she returns to Earth). She declares: "I'm done with just driving. Let's go home." In addition, the continuous thirteen-minute shot that opens the film, together with its breathtaking use of three-dimensional cinematography, reinforces the connection between Ryan's uninterrupted driving and space travel. Whereas the experience of zero gravity may cause Ryan temporary queasiness, the spectator is presumably meant to experience this extended weightlessness as pleasurable in order to mark a dramatic contrast with the cinematographic vertigo subsequent to the destruction of the Russian satellite, whose debris impact sends both the characters and the viewer spinning out of control. The film thus immediately pulls us as spectators into the scene in order that we might glide along with Ryan, adrift between the twin forces of a literal and allegorical loss of gravity.

That Ryan exhibits a melancholia of *zero gravity*—that is, a desire to escape both the weight of her daughter's death and the gravitational pull that drew her daughter toward death—is not to say that she consciously holds the physical laws of attraction (or more accurately, the pull of space curvature) as culpable for her daughter's death. Yet the senseless circumstances that led to the loss are clearly a source of great psychic pain. "Stupidest thing—a school trip to the swimming pool. She was playing tag—she slipped, hit her head, and that was it." There is no agency accountable for the death other than the physical laws of the universe, though Ryan likely feels a lingering guilt for not having been there to catch her fall, a parental "failure" that is reenacted when Matt too is lost to the forces of gravity, leaving her to plead "I had you! I had you."

While zero gravity may provide a means of escape, the aptly named Stone must eventually fall to Earth; or rather, she must come to learn that

Figure 3. Ryan grounded in *Gravity*.

her continued survival depends on her ceasing to defy gravity, thereby allowing herself to be drawn back toward the ground "responsible" for her daughter's death. The film concludes when Ryan crash-lands just off the coast of an undisclosed location. She struggles to swim to shore and collapses face down in the sand, weakened by the effects of prolonged exposure to weightlessness (Figure 3). Whereas her daughter never rose after her fall, Ryan musters enough strength to pull herself up. The film ends when she begins to take a few tentative steps forward, like a child learning to walk again. Ryan may rediscover her footing, but the film concludes without confirmation of her rescue. Before she leaves the escape pod, we hear Houston on the radio asking her to confirm identity and stating that "a rescue mission is on the way," yet we never see any rescuers arrive. Indeed, the film ends where so many other lost-at-sea narratives begin: with the lone protagonist crawling onto a deserted beach uncertain of ever being found. Presumably she will be saved, but the film portrays her reunion with humanity as secondary to the salvation that being grounded provides.

The Eclipse of the Subject

Is OOO's defiance of gravity any more sustainable than Ryan's? Rhetorically, OOO advances its indictment of so-called correlationism by producing litanies of assorted objects, as if by their sheer repetitive reproduction the human might be roused from its "correlationist slumber."[9] As Harman argues, the definition of "object" must be broadly construed to "include those entities that are neither physical nor even real."[10] The "stars" of Harman's three-ring circus include clowns, diamonds, rope, neutrons, armies, monsters, square circles, cotton, fire, pixies, nymphs, utopias, sailboats, and atoms—a surfeit of objects whose evident dissimilarity provides an opportunity to stress their common thinginess. Similar to Harman, Bogost questions the undue value and influence afforded to select things by virtue of how they "relate to human productivity, culture, and politics."[11] He thus promotes a "tiny ontology," which he likens to a black hole's capacity both for infinite density and infinite expansion.[12] Just as some physicists suggest that the infinitesimal singularity at the center of a black hole might contain an entire universe, "small" objects are no less important than large ones. Size apparently doesn't matter for OOO (unlike its XXX counterpart).

OOO's insistence that "subjects are objects *among* objects, rather than constant points of reference related to all other objects," marks precisely the distance it seeks to take from phenomenology, which also promises to return us "to the things themselves," but does so in a manner that paradoxically parenthesizes the objective world rather than the subjective one.[13] As I argued in chapter 1, the Husserlian transcendental reduction suspends the world as the condition of our intentional relation to it. Intentionality is utterly impossible without this quasi-worldly transcendence whereby we find ourselves at the aporetic threshold between inside and outside, as if orbiting along what scientists refer to as the Kármán line: an invisible border that separates Earth's atmosphere from outer space, a boundary that can be witnessed only indirectly by the formation of the Aurora Borealis. In *Gravity*, the spectator briefly views this luminous phenomenon when the camera zooms out from the window of the Soyuz in the wake of Ryan's discovery that the capsule is depleted of fuel, as if dooming her to persist in an utterly "transcendental" orbit far beyond Earth. To be entirely outside the world, in Husserlian terms, is no better than to be fully inside it. Far from eclipsing the world, intentionality requires an intramundane subject who remains forever suspended between an empirical and transcendental "altitude."

Denying the subject any transcendental space, OOO constructs a false choice between transcendence and immanence. This philosophy of pure immanence thus fails to take into account the subject's liminality. The specter of solipsism is often invoked as a menace to be avoided at all costs, a threat to all intersubjectivity, to all social, ethical, and political obligations. Notwithstanding the truly risible view of the world as merely a product of the ego, the ego is always quasi-solipsistic, belonging to what Husserl characterizes more or less interchangeably as "transcendental intersubjectivity" or a "community of monads."[14] This space is not to be understood in terms of an absolute erasure of the world, but rather as the solitude (*solus*) of a self (*ipse*) that is never entirely enclosed nor fully open to the outside. The other exists both within and without us as precisely what will always escape the ego and prohibit its total enislement.

Endorsing a world of subjectless objects, OOO thus commits precisely the error that Husserl cautions us to avoid: it disavows the I's "indeclinability" as the zero point of orientation of the world.[15] OOO responds to phenomenology by reversing its temporary suspension of the world and supplanting it with a permanent and total eclipse of the subject. Yet no subject can perceive and experience the world except by inhabiting a virtual space of exception, of semi-exteriority and privilege. Far from fully enclosing the subject, the transcendental reduction disallows any originary relation to the world that would efface the distinction between self and other. Whereas the transcendental reduction's apparent solipsism forestalls the reduction of otherness to the same, OOO's plane of immanence utterly flattens differences, thus depriving the quasi-transcendental subject of relating to alterity altogether.

Despite railing against philosophical correlationism, Harman surprisingly claims not to accept Meillassoux's wholesale rejection of Kant. Instead, he maintains that Kantian finitude can be "retained but also expanded well beyond the realm of human-world interaction."[16] Harman accordingly has developed a theory of object withdrawal, which asserts that relations *between* insentient things are themselves "haunted by the inaccessibility of the thing-in-itself."[17] While he frames object withdrawal as a radicalization of Kant, the more direct antecedent is Husserlian appresentation whereby our adumbrations of objects remain perceptually inexhaustible. For Harman, "inanimate collisions [between two billiard balls, for example] must be treated in exactly the same way as human perceptions."[18] Objects interact indirectly and incompletely; they never fully "know" one another. He does not deny that "human experience is rather different from inanimate contact, and presumably richer and more complex."[19] Yet he nevertheless maintains that the difference between human relations with insentient objects and their relations with one another is simply "a matter of degree" rather than of kind.[20] Nonhuman things clearly display agency, at least broadly understood as an inherent capacity for action. Spinoza, for instance, argued that a falling stone "is endeavouring, as far as in it lies, to continue in its motion."[21] But does a stone have intentions? Does it have perceptions and experiences such that the difference between our relation to objects and their relation to one another is truly a matter of degree?

Given that Harman accepts Kantian finitude, one can only be puzzled by the allegation of correlationism ("a pejorative term deserving of widespread use") repeatedly leveled at poststructuralist theory: "Rather than direct discussion of the world in itself," the linguistic turn adopts the "robotic post-Kantian gesture" of viewing language as "the condition of access to the world."[22] Yet Harman remains a "closet" correlationist against his stated intentions insofar as he imagines objects as withdrawing from relation with one another. For how else can he claim to expand Kantian finitude except by taking human-world finitude as the prototype for inter-object withdrawal? To erase the distinction between differences of degree and differences of kind is to imply that inanimate things phenomenalize one another. Whatever one might conclude about the proposition that inanimate objects "perceive" one another, suffice it to say that the theorization of inter-object intentionality presupposes a prior conception of human intentionality. Further evidence of this retreat from the antirelational thesis can be found in Harman's declaration that intentionality is Husserl's "greatest contribution to philosophy."[23] Such praise cannot be reconciled with Harman's rejection of the transcendental reduction on the basis of its alleged idealism because the reduction conditions precisely the move from the natural to the phenomenological, *intentional* attitude that Harman endorses.

If OOO "knows" that relationality is inescapable, then why does it overdramatize the alleged independent, nonrelational reality of objects? Jane Bennett suggests that the stakes of OOO lie both in "the pleasure of iconoclasm" and in its implied ethico-political commitments.[24] That the former takes center stage in OOO, however, is confirmed by the looseness with which OOO invokes ethical and political terms. Whereas a considerable number of posthumanist theorists are animated by a political or ethical imperative to interrogate the human/animal hierarchy on the basis of which the lives of nonhuman animals are deemed trivial and disposable, OOO claims to restrict its concern to the ontological realm. Yet a cryptopolitical and ethical stance pervades OOO, surfacing frequently in inchoate demands for "justice" and "democracy" to combat the human "prejudice" of correlationism. According to Bogost, for instance, animal studies is "not posthumanist enough" because its zoocentrism disregards plants, fungi, and bacteria.[25] Harman similarly declares that "no philosophy does justice to the world unless it treats all relations as equally relations."[26] Rather than articulate the ethical stakes of OOO, however, he tacitly appeals to readers primed to view all hierarchies as intrinsically pernicious, as if the human/object hierarchy demanded a response equal in urgency to the fight against speciesism, racism, sexism, and other social ills. OOO thus does not tolerate any hierarchy of hierarchies. It not only treats all hierarchies as equal, but conflates them with difference as such, as if unknowingly employing a caricature of deconstruction. As Derrida insists in *The Animal That Therefore I Am*, the general singular "Animal" effaces "the infinite space that separates the lizard from the dog, the protozoon from the dolphin, the shark from the lamb, the parrot from the chimpanzee, the camel from the eagle, the squirrel from the tiger, the elephant from the cat, the ant from the silkworm, or the hedgehog from the echidna."[27] To erase the distinction between humans, animals, and things is to supplant a hierarchy with a specious homogeneity, as if the general singular "Object" has come to supplant "Animal" as the catchall category in which everything becomes interchangeable with everything else. That we know that a hedgehog is not the same as an echidna, however, belies any pretention to a noncorrelationist perspective, which at bottom appeals to nothing more than an impossible view from nowhere.

OOO thus relies on a form of philosophical bait-and-switch that deploys a liberal pluralist language of justice, equality, and inclusivity only to denude these terms of their ethico-political import. This tactic is nowhere more evident than in Levi Bryant's *The Democracy of Objects*. We quickly learn that this title does not reflect any "*political* thesis to the effect that all objects ought to be treated equally or that all objects ought to participate in human affairs. The democracy of objects is the *ontological* thesis that all objects, as Ian Bogost has so nicely put it, equally exist while they do not exist equally."[28] This characterization begs the question as to the purpose of employing a political concept only to deny its political valence. As Peter Gratton observes, moreover, Bryant's claim that OOO does not exclude humans, but only renders them objects among objects "risks an object-oriented political correctness," such that (paraphrasing Arendt) if "everything is due justice and respect, then nothing is."[29] Bogost likewise invokes parity in a nonpolitical sense when he states that his conception of object equality holds only that objects are irreducibly different from one another: "The funeral pyre is not the same as the aardvark; the porceletta shell is not equivalent to the rugby ball."[30] Yet if that is the case, then surely we must conclude that, stripped of any political and ethical commitments, Bogost's maxim that Bryant cites so admiringly says nothing more than *many different objects exist.*

While OOO does not explicitly advance an object-oriented ethics, it nevertheless shares with Emmanuel Levinas a conception of the absolute designed to sidestep the quasi-transcendental ego. Refusing to see the other as an alter ego insofar as this formulation presupposes the primacy of *my* ego, of an "I" who subordinates otherness to itself, Levinas insists that the other is absolutely other to the same.[31] He thus rejects the Husserlian view that the self remains the indeclinable zero point of any relation to alterity. Harman explicitly acknowledges an affinity with Levinas's desire "to go beyond relationality altogether."[32] As Derrida argues in his critique of Levinas in "Violence and Metaphysics," however, the refusal to recognize the ego's quasi-transcendentality by locating the ego "entirely in the world and not, as ego, the origin of the world" is "the very gesture of all violence. . . . The nonrelation of the same to the other . . . is pure violence."[33] To absolve the other of any relation to ipseity is to utterly evacuate it. Other must mean *other than* me, and no amount of openness to alterity can surmount this relational violence. The point is not that OOO ultimately does violence to objects in the same manner that the assertion of the other ego's absolute alterity inadvertently commits violence. Rather, OOO's insistence on the human's nonrelation to objects—their radical independence from us—presupposes precisely the relationality it aims to escape.

Bennett suggests that the implicit ethical commitments of OOO are also manifest in its critique of human hubris.[34] Bryant, for instance, contends that renouncing human subjects as "monarchs" among beings will lead to "a genuinely post-humanist, realist ontology."[35] Harman similarly observes: "However interesting we humans may be to ourselves, we are apparently in no way central to the cosmic drama, marooned as we are on an average sized planet near a mediocre sun, and confined to a tiny portion of the history of the universe. All these apparent facts are sacrificed in the name of superior rigor, by Kant's Copernican philosophy and its successors."[36] Setting aside for the moment the question of whether Kantian finitude ought to be understood as "sacrificing" the ostensible fact of human mediocrity and triviality, we would do well to scrutinize how this critique expels the human only to usher it in through the backdoor. The desire to sacrifice narcissism altogether, to negate self-interest in the name of the allegedly nonrelational, sovereign agency of things, can only reinscribe human sovereignty. As we have seen previously, the sovereign is by definition all alone, absolute and absolved from any relation, even though this pure form of sovereignty never exists in fact, even though sovereignty is always already shared. As soon as sovereignty gives reasons to justify its absolute force, sovereignty fractures from within. That the nonrelational is only another name for sovereignty suggests that the purported sovereignty of things is but the photographic negative of an auto-affective *desire* for sovereignty that has been disavowed and projected onto things themselves. OOO knows what Viney knows in "The Dumb Witness": the voice that keeps silent retains its sovereignty. The paraleptic, "silent" sovereignty of OOO incessantly talks about not talking about humans, thus ironically fostering an ultrahumanism that bolsters human dominance. Whereas Viney employs silence as a mode of resistance to slavery—an institution that authorizes her ontological and juridical oscillation between animal and thing—OOO uses silence to consolidate human power in the name of "real" things.[37] This "strongest" of all posthumanist theories ultimately discloses itself as the "weakest"—that is, the most conventionally human-ist—not despite but because of the authority it grants itself to exceed the human, an authority that thereby remains paradoxically enslaved to ipseity, to what Derrida describes as the "I can" that conditions freedom as such.[38] As Michael Naas notes, "every form of sovereignty" is "a phantasm, and every phantasm a phantasm of sovereignty."[39] That the phantasm of human sovereignty does not "exist beyond its appearance" ensures that "it is per-petually resurrected because of this nonexistence."[40] This ceaseless resur-rection must be matched with a ceaseless deconstruction that is no match, as it were, for the phantasm it seeks to oppose. It must be a nonreactive weak force that recognizes its own phantasmatic power. The critique of sovereign power must avoid becoming a counter-sovereignty; it must cul-tivate an agency that is no more indemnified against its own ruin than any other power.

The full and final displacement of human exceptionalism is not a pre-sentable, achievable outcome: not a done-and-dusted "deconstruction." The project of decentering the human belongs to the urgency of the here and now even as it remains eternally incomplete. By contrast, OOO seeks to leave all other theories in the dust as it impatiently races ahead toward a "non-sovereignty" that turns out to be more sovereign than ever, despite rebranding itself as a tiny ontology, a motto whose marketing potential Bogost foresees as arising from its simplicity: "The embroiderable short-hand for tiny ontology might read simply, *is*. . . . Theories of being tend to be grandiose, but they need not be, because being is simple. Simple enough that it could be rendered via screen print on a trucker's cap."[41] This hipster nanosovereignty is asymptotically related to a hyperbolic sovereignty that relentlessly ups the ante, that reaches for a sovereignty beyond sovereignty by claiming to outstrip all others.[42] Nanosovereignty declares its diminutive status, its apparent modesty, all the more to conceal the grandiosity of its bid to surpass correlationism, to transcend the aporia of Kantian finitude by insisting on the absolute sovereignty of things. OOO thus grants objects a power to withdraw absolutely both from one other and from the human—a sovereign capacity that is not ours to give.

The Next Big Thing

While OOO's nonrelational melancholy disavows the loss of an unmediated relation to the world, it also betrays many of the characteristics that Freud associates with mania, a condition whose histrionics and triumphalism are symptomatologically opposed to melancholia but that nevertheless, he maintains, belongs to the same malady: "In melancholia the ego has succumbed to the complex whereas in mania it has mastered it or pushed it aside. . . . The manic subject plainly demonstrates his liberation from the object which was the cause of his suffering" by hunting "like a ravenously hungry man for new object-cathexes."[43] In an effort to manage the loss of the thing-in-itself, OOO grasps indiscriminately at new objects whose equal ontological status is declared by obeying an entirely arbitrary, unmotivated classificatory procedure.[44] Harman suggests that philosophy ought to recover "its original character as Eros" by adopting an "erotic model" as "the basic aspiration of object-oriented philosophy: the only way, in the present philosophical climate, to do justice to the *love* of wisdom that makes no claim to be an actual wisdom."[45] OOO makes no claim to produce wisdom because "the real is something that cannot be known, only loved."[46] Yet beneath this professed love of objects subsists an utter indifference to them. As Christopher Norris remarks, "there is not much point in continually reeling off great lists of wildly assorted objects if the upshot is merely to remark on their extreme diversity, or irreducible thinginess, without (as it seems) much interest in just what makes them the way they are."[47] Indeed, to love every object equally is to love no object, save for the Kantian thing-in-itself, which is the only object to which OOO can openly declare its love, the object that Big Bad Kant, the "common enemy" of speculative realism, has refused us.[48] Whereas Freud observes that in mania "what the ego has surmounted and what it is triumphing over remain hidden from it," here it seems that the object is hiding in plain sight on the pages of every book and article written by an object-oriented ontologist or a speculative realist.[49] Freud suggests that the manic subject "must have got over the loss of the object (or its mourning over the loss, or perhaps the object itself)," which releases an abundance of psychic energy "available for numerous applications and possibilities of discharge."[50] Yet if the voracious appetite for new objects characteristic of mania belongs to the same complex as melancholia, then it remains doubtful that this enthusiasm can be read as heralding the relinquishment of the object.

OOO exhibits its hunger for novelty not only by expressing boredom and impatience with correlationism, but also by euphorically declaring its own innovative promise. Psychoanalyst Adam Phillips has suggested that boredom shares with melancholia a sense of unnamable loss; boredom amounts to a "defense against waiting" that results in two contradictory attitudes: "There is something I desire, and there is nothing I desire. . . . In boredom there is the lure of a possible object of desire, and the lure of the escape from desire, of its meaninglessness."[51] OOO's indifferent litanizing suggests a comparable split attitude: "I desire every object; I desire no object in particular." This conception of boredom as a desire for all and no objects also sheds light on OOO itself as an object of theoretical attention. As Derrida observes, the establishment of new theoretical paradigms is often characterized by a preoccupation with novelty that says "shove off" [*pousse-toi de là que je m'y mette*] to previous theories according to a tactic that "reveals too much impatience, juvenile jubilation, or mechanical eagerness."[52] Consider the elation that Morton evinces when he describes OOO as a "cool flavor [that] fizzes with the future—the bliss of new thinking."[53] Later when he asks whether "reality . . . shame[s] 'linguistic turn' self-absorption with a gigantic, massively distributed raspberry," he displays precisely the impatience that Derrida describes.[54] Similarly, Bryant declares that "object-oriented ontologists have grown weary of a debate that has gone on for over two centuries, believe that the possible variations of these positions have exhausted themselves, and want to move on to talking about other things," as if fatigue is all the justification one needs for pronouncing correlationism's demise.[55]

As I noted in the introduction, this exhaustion and impatience correlates with increased publication pressures and decreased academic employment. Given the corporate culture of modern universities, their obsession with market positioning and incessant rebranding, is it any surprise that some academics would follow suit? Harman undoubtedly seeks to position his brand when he heralds OOO as "the next big thing," as if to reify OOO into a merchandisable logo, one that should be embraced (allegedly) "not for the sake of earning social capital and a with-it image, but because any theoretical content eventually reaches a point where it is no longer liberating." He continues, "From time to time something new is needed to awaken us from various dogmatic slumbers. Properly pursued, the search for 'the next big thing' is not a form of hip posturing or capitalist commodification, but of hope."[56] How is its proper pursuit to be defined? Why does the search for novelty equate with hopefulness? That the purported liberatory power of any theory is *not* the measure of its descriptive truth or untruth, its relevance or irrelevance, suggests that this assertion of correlationism's sell-by date accedes precisely to the hip posturing that Harman disavows. Exasperated with feeling "trapped in the correlational circle," Harman attaches to OOO as a fetishized object, one whose trinity of Os fittingly emblematize both hollowness and hermetic closure, as if OOO ultimately wraps itself around its own empty center, "correlating" only with itself.[57] If everything is an object in the world of OOO, and no object is more worthy than any other, then how ought we to respond to a theoretical approach that shamelessly presents itself as the next big thing? Is this not simply "a shape to fill a lack," to borrow from William Faulkner, a stopgap measure that revolves back to itself through the self-same movement that enjoins the rest of us to get over ourselves and affirm the lives of things?[58]

Rather than seek to puncture the alleged correlational circle, to transcend its limits by insisting on the horizontal immanency of human, animal, and thing, perhaps we might engage an alternative reading of the iconographic O around which this decentering mission incessantly circles. The horizontal arrangement of these Os inscribes a series of separate but adjacent spheres, implying a constellation of fully independent worlds. Notwithstanding the melancholic detachment that this nonrelationality implies, however, this series might be reinscribed as a chain of overlapping Os, like intersecting, immanently transcendent planetary orbits, interlaced yet irreducibly distant. These Os would be reminiscent of Friday's "tightly packed together" rows of the letter *o*, whose graphic sphericity nevertheless implies a certain proximity of subjective worlds, touching one another without ever becoming fully identical.[59]

We Are All Ptolemaists

Although Meillassoux's rhetoric is far more tempered than that of his OOO followers, his argument is not entirely devoid of histrionics. After all, his philosophical project alleges that the "Kantian catastrophe" has deceived philosophers into accepting the finitude of human knowledge.[60] Meillassoux thus audaciously seeks to set philosophy straight, to reorient it away from the wrong turn of "transcendental idealism," to take the path of the *"speculative materialism"* toward which philosophy should have turned over two centuries ago.[61] For Meillassoux, "ancestrality," defined as "any reality anterior to the emergence of the human species," belies the Kantian insistence that "cognition reaches appearances only, leaving the thing in itself as something actual for itself but uncognized by us."[62] According to Meillassoux, the correlationist cannot accept the "ultimate meaning" of a statement such as "Event Y occurred x number of years before the emergence of humans."[63] This ultimate meaning, which he explicitly equates to a literal one, cannot be deepened by the ostensibly more "originary" phenomenological interpretation of the correlationist who, though contesting neither the occurrence of the event nor its priority, "will simply add—perhaps only to himself, but add it he will—something like a simple codicil, always the same one, which he will discretely [*sic*] append to the end of the phrase: event Y occurred x number of years before the commencement of humans—*for humans*."[64] Although Meillassoux names Heidegger and Husserl as inheritors and exacerbators of Kant's disastrous critical turn, for the most part "correlationist" is broadly invoked, allowing the stigma to be worn by virtually anyone who reads the ancestral statement otherwise than purely literally. This rhetorical strategy permits him to claim that the correlationist views the logical priority of givenness, the scientific statement's "deeper, more originary" meaning for the human, as "truly correct."[65] For instance, the correlationist allegedly "cease[s] to believe that the accretion of the earth straightforwardly preceded in time the emergence of humanity."[66] Yet does the correlationist in fact subscribe to such a zero-sum game of literality versus figurality: *of* us versus *for* us? Husserl, for instance, critiqued the modern scientific tendency to reduce reality to objective description, but he nevertheless did not view science and phenomenology as mutually exclusive. He thus saw no contradiction in placing Galileo "at the top of the list of the greatest discoverers of modern times," even while charging him with substituting a pure geometry of ideal shapes (the perfect circle, the perfect triangle, etc.) for our prescientific world of lived experience, the *life-world* that gave rise to the practice of imperfect geometric measurement (land surveying, building construction, and astronomy). The effort to exhaustively mathematize nature fails to inquire back into the lifeworld, which is the condition of possibility for geometrical idealization.[67] The reduction of reality to geometric ideality thus risks estranging and alienating us from the lifeworld. To be sure, Husserl characterizes the lifeworld as "the only real world," but this does not map so easily onto Meillassoux's claim that the correlationist views phenomenological givenness as "truly correct" in an absolutist sense. Indeed, Husserl adopts a conciliatory approach that Meillassoux chastises precisely for not being absolutist enough. Hence, the alleged intransigence in asserting givenness *for us* as truly correct is difficult to square with the inconsistency of which Meillassoux also accuses the correlationist: "A consistent correlationist should stop 'compromising' with science and stop believing that she can reconcile the two levels of meaning without undermining the content of the scientific statement she claims to be dealing with. There is no possible compromise between the correlation and the arche-fossil: once one has acknowledged one, one has thereby disqualified the other."[68]

On the one hand, the correlationist is rebuked for believing in the deeper meaning of the ancestral statement, which is to say for not truly believing its literal truth; on the other hand, the correlationist is scorned for not holding fast enough to this nonbelief, for believing that he can compromise between belief and nonbelief. Yet if compromise is the correlationist's default position—that is, if the correlationist is inconsistent by not subscribing to correlationism/noncorrelationism as a zero-sum game, then Meillassoux can only attribute to the correlationist a belief in the phenomenological attitude's true correctness by inconsistently characterizing the correlationist as being at once too consistent and too inconsistent, too dogmatic yet not dogmatic enough. By Meillassoux's own admission, correlationists do not speak of the phenomenological attitude as the truly correct interpretation (they are far too compromising). Hence, the portrayal of correlationists as dogmatically asserting their perspective as deeper, more originary, and truer reflects the position that Meillassoux *wants* correlationists to hold, a position that constitutes nothing less than the specular image of his own uncompromising dogmatism: either you are with the correlationists or you are against them.

This stubborn insistence on literal reading, however, proves difficult to sustain when Meillassoux shifts his focus to the Copernican revolution, whose meaning he locates not in the scientific fact of heliocentrism, but in what this fact says about scientific reality's indifference to human observation: "*The Galilean-Copernican revolution has no other meaning* than that of the paradoxical unveiling of thought's capacity to think what there is whether thought exists or not [my emphasis]."[69] Meillassoux thus privileges an interpretation of the Copernican turn that is said to be more fundamental, a meaning more originary and deeper than most scientists and lay people would typically be inclined to consider when the name Copernicus is mentioned, despite having flatly dismissed any level of meaning other than a flat one. If the Copernican revolution truly meant only that "what is mathematizable cannot be reduced to a correlate of thought," then we would be forced to accept the rather bizarre conclusion that the Copernican revolution no longer means that the earth revolves around the sun.[70] It thus turns out that the "true" meaning that Meillassoux assigns to the Copernican revolution depends on him adding a correlationist codicil: scientific truth cannot be reduced to a correlate of thought—for Meillassoux.

While he emphasizes the "deeper" discovery of the world's radical indifference to human existence that Copernicus ostensibly brings to light, he nevertheless falls back on the more literal fact of astronomical displacement when he suggests that Kant's revolution is Ptolemaic because it asserts "not that the observer whom we thought was motionless is in fact orbiting around the observed sun, but on the contrary, that the subject is central to the process of knowledge."[71] As this passage clearly demonstrates, the accusation of Kant's Ptolemaism relies on two related but different analogies: (1) Kant is Ptolemaic because he inscribes the human as the locus of epistemological access onto a universe whose indifference to human thought is thoroughly demonstrated by the total mathematization of celestial movement; (2) Kant is Ptolemaic because he reinscribes the epistemological centrality of the human as a direct reaction against the cosmological decentering wrought by Copernicus. The first analogy stresses the irreducibility of mathematical truth, the second, the discovery of heliocentrism itself; Kant disavows the astronomical observer's orbit by installing the transcendental subject as an immovable center. Heliocentrism shares the same evidentiary purpose—to refute correlationism—as do ancestral statements regarding the age of the universe, the accretion of Earth, the origin of life, and the origin of human life. Yet heliocentrism bears a rhetorical significance quite unlike these other scientific truths. The literality of cosmological decentering is invoked in order to indicate a *figurative* recentering of the human as a subject of knowledge. Ancestral statements do not supply Meillassoux with the same rhetorical *coup de grace* against Kantianism. Indeed, they are said to signify only literally. Yet it is precisely the figurative sense of geocentrism that gives his critique of Kant such rhetorical force. It paints Kant as regressive and reactionary, as if to imply a hidden rapprochement between an outmoded view of the physical world and the critical turn.

This equation of Kantian epistemology with a figurative geocentrism is also problematic from a historical perspective because it presumes that the center was viewed in the pre-Copernican world as an unequivocally desirable position when the reverse is actually true. Meillassoux is not unaware of this fact, though it is curiously buried in one of his footnotes: "The end of Ptolemaic astronomy does not mean that humanity felt itself humiliated because it could no longer think of itself as occupying the centre of the world. In actuality, the centrality of the Earth was then considered to be a shameful rather than a glorious position in the cosmos—a kind of sublunary rubbish dump."[72] He goes on to cite historian Rémi Brague, who provides a wealth of evidence contravening Freud's famous claim that Copernicus inflicted human narcissism with a "wounding blow."[73] Maimonides, for instance, expressed a commonly held medieval view when he wrote that "whenever the bodies are near the center, they grow dimmer and their substance coarser, and their motion becomes more difficult, while their light and transparency disappear because of their distance from the noble, luminous, transparent, moving, subtle, and simple body—I mean heaven."[74] Readers of Dante will likewise recall that the *Divine Comedy* locates hell at the center of the earth: "In the world of sense we can perceive / That evermore the circles are diviner / As they are from the centre more remote."[75] These premodern assessments of Earth's insignificance resonate with Harman's remark that we live "on an average sized planet near a mediocre sun," yet in so doing they underscore that the recognition of earthly, human mediocrity is hardly a novel insight. Moreover, if the center of the universe was deemed a shameful position prior to Copernicus, then how can Kant be construed as reacting against Earth's displacement? It makes little sense to charge Kant with enacting a Ptolemaic counterrevolution if Copernicus did not oust the human from its privileged position within the universe. Dennis Danielson argues that this cliché developed in the seventeenth century once the sun was established as the center and this centrality became associated with specialness: "We anachronistically read the physical center's post-Copernican excellence back into the pre-Copernican world picture—and so turn it upside down" (1033). Galileo, for instance, regarded heliocentrism as a promotion rather than a demotion: "As to the earth, we are trying to make it more noble and more perfect insofar as we strive to make it similar to the heavenly bodies and in a sense to place it in heaven, from which your philosophers have banished it."[76] This double movement of demotion/promotion reaches its apex in the self-congratulatory tenor of speculative realism, which divides the world into two species of beings, correlationists and noncorrelationists: the latter "blessed" with an exceptional capacity to recognize the human's nonexceptionality.

The Orbital "I"

That Meillassoux finds himself suspended between the literal and figurative meanings of geocentrism is altogether unsurprising giving that the historical event of Copernicanism involves precisely the problem of a turn that *turns on* any effort to ground its meaning either literally or figuratively. The Copernican turn is not simply one historical revolution among others. Unlike sundry political upheavals, the Copernican event quite literally involves a turn: the revolution of the earth around the sun. This revolution is denoted by the title of Copernicus's sixteenth-century treatise, *On the Revolutions of the Celestial Spheres*. Prior to its recognition as a revolutionary, transformative shift in scientific history, the Copernican revolution named precisely the historical moment when the human turned its gaze away from celestial objects and revolved back toward itself precisely in order to displace our previous belief in Earth's nonorbiting, immovable, centrality. As Kant remarked in the preface to the second edition of the *Critique*, heliocentrism "would have remained forever undiscovered if Copernicus had not ventured, in a manner contradictory to the senses yet true, to seek for the observed movements not in the objects of the heavens but in their observer."[77] While Kant employs the word "revolution" to construct an analogy between Copernican science and the critical turn in philosophy, he also uses the term according to its astronomical sense, as when he writes that Copernicus "made the observer revolve and left the stars at rest."[78] As Hannah Arendt notes, "revolution" began to assume a political valence in the seventeenth century that associated it with the restoration of a previously established order.[79] The term *revolution* acquired its modern association with novelty during the course of the French and American revolutions in which the latent astronomical connotation of turning back yielded to a sense of *irresistibility*, an impulse itself inherited from the astronomical notion that the stars follow a preordained path. The principle of irresistibility was thus transplanted from the heavens to the earthly realm as characteristic of the unstoppable momentum of political transformation.[80]

Alleging that Kant instigated a Ptolemaic counterrevolution, Meillassoux is clearly employing a modern conception of revolution as transformation. While the phrase "Copernican revolution" may have served philosophers for two centuries as a convenient shorthand to describe the critical turn, it does not do justice to a philosophical reorientation that occurred at a historical moment when the term *revolution* was only just beginning to acquire its modern sense, a period in which the distinction between *revolution* and *counterrevolution* was utterly in flux. This is not to suggest that modernity has fully stabilized the meaning of *revolution*, in which case the knowing subject's centrality could also be fully grounded. On the contrary, correlationism unsettles precisely the distinction between periphery and center, turn and return, novelty and restoration. The so-called correlational subject is thus better characterized as an orbital "I" who perceives the world from the groundless ground of an earth that feels stationary even as it moves. As Husserl puts it, the earth is the "originary ark" that "makes possible in the first place the sense of all motion and all rest as mode of one motion. But its rest is not a mode of motion."[81] Husserl is not simply rehearsing the scientific truism that we are unable to perceive directly the earth's rotation and orbit because we are anchored to its movement. On the contrary, the originary ark functions as a figure for the quasi-transcendental ego's "stationary" status. Just as we experience the earth as a ground-body for our perception of other astronomical bodies, the ego is not merely one "planet" among others, but is the primordial homeland for our relation to otherness: "The earth itself does not move and does not rest; only in relation to it are movement and rest given as having their sense of movement and rest."[82] Drawing from the double meaning of orbit as both a stable anatomical center of visual perception (i.e., the eye socket) and a figure of astronomical movement, the orbital "I" affirms an insuperable "Ptolemaism" as the condition of possibility for Copernicanism.

Although we may perceive the center as a place of security and safety, it also represents a particularly vulnerable position to occupy, the point from which one cannot take any distance. As Husserl remarks, "I do not move away; I stand still or go; thus my flesh is the center and the bodies at rest and moving are around me, and I have a ground that does not move."[83] The perils of immobility are manifest in *Melancholia* when Leo says to Justine, "Dad says there's nothing to do then [if the planets collide], nowhere to hide." The wire apparatus that Leo constructs earlier in the film in order to measure the distance and size of the approaching planet serves to remind Claire of the vulnerability of being inescapably fixed. Similar to the "magic cave," a tepee-like frame of sticks that Justine tells Leo will protect them from Melancholia's impact, Leo's apparatus is a decidedly low-tech instrument: nothing more than a wire ring attached to a stick. When held up to the sky, the device permits the viewer to ascertain Melancholia's movement toward and away from Earth as the planet's perimeter shifts inside and outside the circular frame. Initially comforted when she observes the planet decreasing in size, Claire suddenly panics when its perimeter expands beyond the edges of the wire, indicating that the planet is moving closer to Earth. Unlike a telescope—which allows one to manipulate size and distance, thus giving the illusion not simply of bringing the object closer to the viewer but also of the spectator moving closer to the object, as if being launched into space—this anti-telescope dooms Claire to a helpless immobility. The center names a space of apparent exception that paradoxically makes itself vulnerable to annihilation precisely by being what it is: that is, by being the nondisplaceable ground zero for its own deracination.

Bearing the World

While Meillassoux focuses largely on the problem of anteriority in relation to the temporal discrepancy between thinking and a world indifferent to human thought, he briefly notes that this hiatus also concerns "statements about possible events that are ulterior to the extinction of the human species."[84] Such ulteriority is a central concern in *Melancholia* when Claire and Justine ponder the impending arrival of the eponymous rogue planet. As they discuss the possibility that the planet might collide with Earth, despite scientific calculations to the contrary, they adopt contrasting attitudes toward the prospect of annihilation:

> JUSTINE: The Earth is evil. We don't need to grieve for it.
> CLAIRE: What?
> JUSTINE: Nobody will miss it.
> CLAIRE: But where would Leo grow up?
> JUSTINE: All I know is, life on Earth is evil.
> CLAIRE: There may be life somewhere else.
> JUSTINE: But there isn't.

Justine's assertion that no one will grieve the loss of planet Earth derives from two related, yet not entirely identical beliefs: "the Earth is evil" and "*life* on Earth is evil." That "no one will miss it" is quite literally true from a noncorrelationist perspective because the destruction of Earth would entail the absolute erasure of any witness who could grieve its loss. When Claire responds first by asking where Leo would grow up and then by pondering whether life might exist elsewhere in the universe, she commits the correlationist "sin" of imagining the ulteriority of life *for* her. Meillassoux would no doubt take Claire to task for her "Ptolemaic" reinscription of herself at the center of a world that would no longer exist. Claire's thoughts of total destruction do indeed orbit back toward the world that she imagines as no longer existing, but this revolution in thought, this turning back while turning away, simply describes the ordinary Ptolemaic-Copernicanism (or Copernican-Ptolemaism) fundamental to being an intentional subject. The sun does not literally rise and set *for* us; the sun is not significantly closer to us in the summer than in the winter (Earth's orbit is only slightly elliptical). Schoolchildren are disabused of these correlationist misperceptions early on. That we cannot directly perceive our celestial orbit anymore than we can witness anterior or ulterior events does not mean, however, that we are obliged to dispense with figurative language altogether. The question is thus not Ptolemaism versus non-Ptolemaism. There is no one Ptolemaism. There are only *Ptolemaisms* with stronger and weaker gravitational pulls. And what gravitational pull could be stronger than the one emitted by a philosopher who insists on a literality that arrests meaning at its alleged core? What could be more Ptolemaic than a fixation with fixity that seeks to stay metaphor's vehicular drive to bear meaning from one place to another, to moor meaning to an ultimate center around which orbit a number of supplementary figurative effects, like so many satellites of signification?

If Meillassoux's linguistic Ptolemaism requires that language not get carried away, that it remain tethered to an ultimate, literal ground that forever revolves on its own axis, then what are we to make of Justine and Claire's Ptolemaisms in terms of their respective strengths and weaknesses? Justine's gravitational attraction would seem to be the weaker one insofar as it exhibits a degree of hospitality toward an unknowable future quite unlike the increased anxiety and stress that Claire experiences as the planet draws nigh. In the film's Wagnerian overture, for instance, Melancholia's rival gravitational and electromagnetic influences cause St. Elmo's fire to radiate upward from Justine's fingertips. Justine gazes at this luminous phenomenon with calm fascination, as if Earth's gravitational waning lightens the ponderous effects of her psychological depression. She displays similar serenity when she stands with her arms extended, Christlike, while white butterflies float upward toward the heavens, implying a shared pleasure in a liberating buoyancy that welcomes Melancholia literally with outstretched arms (Figure 4). This image anticipates a scene of comparable self-abandonment that occurs later in the film when Justine lies naked on a riverbank at night, allowing herself to bathe in the light of Melancholia, which has become something like Earth's second moon. Justine stares intensely at the luminous blue planet in a manner that recalls an earlier scene when the wedding revelers launch a number of helium lanterns into the night sky. Justine leans forward to observe the lanterns' ascension through John's telescope. The shot cuts to her telescopic perspective

Figure 4. Justine welcoming Melancholia.

of these man-made celestial objects, before she pulls back from the telescope, her eyes closed. The scene once again cuts to a shot of the heavens,
but this time the lanterns are replaced with her mind's eye view of a suc-
cession of breathtaking supernovae. As with the butterflies and the St.
Elmo's fire, her attraction to the lanterns and other celestial objects implies
an affinity with their rootlessness and errancy.

That Justine abandons herself to a looming planet that appears to be
gaseous further suggests that she eschews the reassurance provided by
Earth's terra firma. While all the gas planets in our solar system are sur-
rounded by rings, however, Melancholia noticeably lacks this characteristic. Together with its blue color, the absence of rings gives Melancholia an
appearance that resembles a larger version of Earth. Does Justine look
away only to see Earth's face staring back at her? What if this other world
amounts to another version of the same, an *alter geo*? What would it mean
if Justine remains earthbound despite her perceptual mobility, that is, if her
regard turns out to be far less errant, and thus far less akin to the planet
toward which she gazes? That Melancholia is said to have been "hiding
behind the sun" underscores that von Trier is drawing from the familiar
science fiction trope of the counter-earth. In Mike Cahill's *Another Earth*
(2011), for example, a planet utterly identical to Earth suddenly appears in
the sky.[85] When Dr. Joan Tallis of SETI (Search for Extraterrestrial Intelligence) attempts to contact "Earth 2" during a live television broadcast,
she is shocked to hear her own voice echoed back to her by a denizen of the
other planet who bears her name and reveals identical biographical details.
This doppelganger effect itself mirrors Andrei Tarkovsky's *Solaris* (1972),
in which the eponymous planet is revealed to constitute a parallel Earth.[86]
As Dr. Snaut declares, "we want to extend the Earth to the borders of the

cosmos. We don't know what to do with other worlds. We don't need other worlds. We need a mirror."[87] Is this not precisely what Husserl means when he tells us that we can travel to the moon while never being able to leave our *ur-Earth* behind? We might consider the moon as a kind of second Earth, but which Earth, he asks, would function as the ground body, the zero point of comparison to the other? The ur-Earth as originary ground could be the moon, if one were to have been born there. Yet the planet Earth as quasi-transcendental *geo* does not move. This *monogeologism* of the other—the trip to the moon or Justine's interplanetary gaze—means that there is always more than one Earth/there is never more than one Earth.

That Justine appears more hospitable toward alterity than her sister thus does not mean that the former escapes epistemological centrism altogether. When she asserts that "life is only on Earth," for instance, and defends her claim on the basis of a clairvoyant ability to "knows things," this rejection of the possibility of extraterrestrial life nevertheless implies that such beings, were they to have existed, would have answered to Claire's demand for a world dependent on being given for some perceiving intelligence. This intuition thus betrays a correlationist perspective notwithstanding her insistence on the absence of life elsewhere in the universe. By contrast, Meillassoux is concerned with the "temporal discrepancy between thinking and being" that emerges once one begins to think the absence of any witness, terrestrial or extraterrestrial, that could exist in the wake of Earth's destruction.[88] Extraterrestrial witnesses would be no less vulnerable to future extinction than earthly beings; thus, the problem of ulteriority would remain relevant for Meillassoux. Yet this thought of ulteriority can still not elude correlationism insofar as it can think its nonexistence only from the perspective of a would-be survivor. When Claire imagines Leo as having survived annihilation, her disavowal only hyperbolizes an irreducible Ptolemaic thinking inherent in apperception as such. As Peggy Kamuf puts it, "the end of all life . . . is precisely what we cannot think except in a mode and as a vestige of survival beyond the annihilation that will therefore not have been total."[89] Survival is irreducible because the "imagination of disaster," in Susan Sontag's phrase, cannot transgress the boundary between life and death, thus compelling us to project ourselves onto the scene of our utter absence.[90]

Psychical Ptolemaism unsurprisingly tugs most strongly when faced with the approach of a rogue planet that weakens the gravitational pull of the earth before obliterating it altogether. In addition to verbally expressing her fear and alarm, Claire is the only character who becomes breathless when the planet approaches, a fact that John attributes to the planet's "taking part of our atmosphere." Later when Justine and Leo are collecting sticks from the nearby woods to construct their "magic cave," the camera pans down toward Justine's boots, under which worms and other insects are seen rising to the surface. While the overture suggests that the insects enjoy Justine's weightlessness, here they seem confused and disoriented by the competing electromagnetic and gravitational impulses, as if attempting to flee the soil in search of safety, just as Claire vainly attempts to escape Earth's destruction by fleeing in a golf cart to a nearby village. Claire's anxiety is unsurprising given that Melancholia heralds not simply the end of *her* world but the end of *the* world because it threatens to eradicate the material foundation for any future telluric life. The overture stresses this fragile footing when it depicts Claire in slow motion carrying Leo across the golf course, returning from her failed escape. This decelerated movement gives the impression of weightlessness, as if gravity no longer holds sway over Claire, each of her steps leaving deeper imprints in the softening soil as we see a flag marking the nineteenth hole unfurl in the background (Figure 5). The weightlessness, footprints, and the flag all evoke the indelible image of the Apollo 11 moonwalk. Whereas NASA footage showing the American flag seemingly blowing in the wind is cited by conspiracy theorists as evidence that the mission was a hoax, von Trier plays his own prank on the viewer by referencing a nonexistent, nineteenth hole that also happens to be a euphemism for the watering hole where one celebrates after a game (Figure 6). When the real-time version of this scene later depicts a volley of hail descending upon Claire, it resembles not only the beans whose triviality she earlier dismissed, but also innumerable golf balls, those man-made projectiles to which large ice pellets are often likened. As von Trier observes, "There's something oddly melancholy about golf courses. They go on forever and, if you take away all the golfers . . . they're amazingly cultivated landscapes. I always loved golf courses and graveyards."[91] Blurring the distinction between the natural and the unnatural, these golf/ice balls fall onto the surface of an artificial green whose cosmic vastness harbors an inordinate number of black holes. Down is up and up is down on this Carrollesque earth-*cum*-outer-space, an infinite expanse soon to be emptied of all humanity.[92]

No longer able to rely on the previously dependable and predictable laws of the universe, Claire finds herself faced with the pressing dilemma not only of how to bear both herself and her child upon a planet that is literally disappearing beneath them, but also of how to bear this world that also bears her. Claire is both carried and carrying. Presumably the cosmos

Figure 5. The nineteenth hole in *Melancholia*.

Figure 6. Apollo 11 moonwalk.

will continue to exist in the wake of Earth's destruction, a universe entirely independent and "indifferent" to her, but for the moment she is still obliged to carry the world as a horizon of perception in anticipation of its utter absence. She has only one world, yet it is not hers. The world has always existed *for* her; it has never existed (solely) for her. This is the performative contradiction that binds her to carry the world even while she is anticipating its disappearance. As Derrida puts it:

> *To carry* now no longer has the meaning of "to comprise" [*comporter*],
> to include, to comprehend in the self, but rather *to carry oneself for bear*
> *oneself toward* [se porter vers] the infinite inappropriability of the other,
> toward the encounter with its absolute transcendence in the very
> inside of me, that is to say, in me outside of me. And I only am, I can
> only be, I *must* only be starting from this strange, dislocated bearing
> of the infinitely other in me. I must carry the other, and carry *you*,
> the other must carry me . . . even there where the world is no longer
> between us or beneath our feet, no longer ensuring mediation or rein-
> forcing a foundation for us.[93]

While Derrida is speaking here of the immanent transcendency of other beings, living or dead, this liminality belongs to our appresentational apprehension of alterity more generally. This directedness *toward*—perhaps both fearful and welcoming—is neither a condition of zero gravity nor a black hole whose event horizon marks the point of no return, the border beyond which otherness is utterly absorbed. Like an astronaut orbiting Earth, atlassing a panoramic view that is perspectivally inexhaustible, we stand on moving ground, bearing ourselves toward a world both in and without us, a world that persists minus our presence but is always *thought* from the zero point of a melancholic gravity that must be more than naught.

CHAPTER 4

Listing Toward Cosmocracy
The Limits of Hospitality

It is by constructing itself on the basis of the vertical
comparison with animality that the cosmopolitical horizon
opens up and its horizontal contract is constituted.

—PETER SZENDY, *Kant in the Land of Extraterrestrials*[1]

In the 1855 preface to *Leaves of Grass*, Walt Whitman writes that the "equable" poet "judges not as the judge judges but as the sun falling round a helpless thing."[2] Illuminating everything without exception, solar judgment appears to beget a capacious and magnanimous ethico-political relation, an illimitable and indiscriminate hospitality that embraces the vulnerable and powerless. According to Jane Bennett, "to go solar, to accept all with equanimity, is to elide the particular hierarchy of values and the particular regime of perception dominant in the culture."[3] Citing Bergson's observation that perception is inherently "subtractive," a procedure for screening out external stimuli to which the subject is indifferent, Bennett suggests that Whitman provides us with an otherwise nonsubtractive, wholly receptive bearing of and toward the world.[4] On this account, solarity is remarkably similar to the plane of immanence championed by object-oriented ontology. Just as OOO promotes a flat ontology that eschews the hierarchical, fractional distribution of subjective intentionality, the sun orients itself toward all things in equal measure.

Solarity finds some precedent in both ecopoetical and queer approaches to Whitman. Jimmie Killingsworth, for instance, argues that Whitman ascertains "the things of nature" in an indirect, nonacquisitive manner, one that affirms the limits of human language to represent otherness.[5] Similarly attuned to the problem of alterity, Peter Coviello maintains that the "embarrassingly immodest . . . confidence with which the poet casts himself in often improbable roles" runs counterpoint to a sustained concern with sympathetic identification.[6] Michael Warner also acknowledges that Whitman "thematizes a modern phenomenology of self everywhere: 'I celebrate myself, and sing myself.'" Yet Warner justifiably takes issue with readers who reduce Whitman's poetic "I" to such declarations of liberal selfhood, arguing that Whitman ultimately engenders "the pragmatics of selfing a mess."[7] In this regard, the absorption of otherness into the self is never fully accomplished. In "To a Stranger," for instance, the speaker addresses someone both anonymous yet known: "You grew up with me, were a boy with me or a girl with me."[8] It remains uncertain whether the poem describes a moment of uncanny misrecognition in which the speaker encounters an other who reminds him of someone previously known or whether the ostensible stranger literally belongs to the poet's past. Whether familiar or strange, this alterity yields to a quintessential Whitmanian merger: "I ate with you and slept with you, your body has become not yours/only nor left my body mine only."[9] On closer inspection, however, this merging does not describe a dialectical synthesis that incorporates alterity: the other's body neither fully belongs to this other nor to the "I" who describes its incomplete arrogation.

As both Warner and Coviello demonstrate, the disarray of self-constitution can be registered only if we dwell with the poetic imbrication of self and other. This nondialectical chiasmus weakens the propensity of the poetic voice to overflow its own boundaries and absorb everything into itself. D. H. Lawrence famously lampooned this absorptive power for its apparent monomaniacal pomposity:

> All that false exuberance. All those lists of things boiled in one pudding-cloth! No, no!
> I don't want all those things inside me, thank you.
> "I reject nothing," says Walt.
> If that is so, one must be a pipe open at both ends, so everything runs through.
>
>   . . .
>
> "I embrace ALL," says Whitman. "I weave all things into myself."
> Do you really! There can't be much left of *you* when you've done.
> When you've cooked the awful pudding of One Identity.
>
>   . . .

> As soon as Walt *knew* a thing, he assumed a One Identity with it. . . .
>
> . . .
>
> This merging, *en masse*, One Identity, Myself monomania was a
> carry-over from the old Love idea.[10]

Lawrence employs *reductio ad absurdum* to show that there is both too
much and too little of Whitman in his poetry. The latter's seeming self-
abandonment merely conceals its arrogance. Of course, the exuberance
that Lawrence himself exhibits here risks imitating the self-glorification
that he wishes to mock. Similar to Lawrence, Doris Sommer argues that
equality is "synonymous with identity" for Whitman, who assumes a self
"at the center of a universe that repeats him endlessly."[11] Undoubtedly a
certain grandness of purpose is not altogether foreign to Whitman. In *Song
of Myself*, he declares: "I know this orbit of mine cannot be swept by a car-
penter's / compass."[12] Several verses later he inscribes his name and identity
directly into the poem: "Walt Whitman, an American, one of the roughs,
a kosmos."[13] A world unto himself, Whitman nevertheless assures us that
he is "no stander above men and women or apart from / them." Yet several
lines before he characterizes the poet in the 1855 preface as one who suspends judgment, the speaker stations himself "high up out of reach," where
"he stands turning a concentrated light . . . he turns the pivot with his fin-
ger . . . he baffles the swiftest runners as he stands and easily overtakes and
envelopes them."[14] The poet *as* sun shines down upon all. Not even the
fastest runners can escape its advancing reach. Nonjudgmental judgment
might appear to equate to a form of incalculable hospitality, an openness
"To You. Whoever you are."[15] Yet what if a judge's calculable discernment
is precisely what is needed to reckon with an utterly incalculable, perhaps
even oppressive, solarity? Can we finally distinguish between a capacious
and a rapacious solarity? Does a nonjudgmental judgment finally amount
to a total absence of judgment without which everything would be help-
lessly exposed, willy-nilly, to the sun's scorching rays? Consider that the
"swiftest runners" might be read as fleeing the sun precisely *because* it has
"baffled" them: that is, confused and obstructed their path. Are they *also*
helpless things, vulnerable and powerless objects of a light that relentlessly
pursues them? If at first glint solar judgment appears to register only magnanimous receptivity, might it not also cast a long shadow of violence that
obscures as much as it illuminates?

Focused on the luminosity of objects, Bennett positions herself squarely
within the Platonic tradition that equates the sun with the good beyond
being, the condition of possibility for sight, the "child of the good," which

is to say knowledge and truth.[16] Solarity thus contrasts with OOO's theory of object withdrawal discussed in chapter 3. OOO seeks to evade the light in a manner that recalls the Levinasian claim that Western metaphysics enacts a phenomenological violence that utterly absorbs alterity: "Light is that through which something is other than myself, but already as if it came from me. The illuminated object is something one encounters, but from the very fact that it is illuminated one encounters it as if it came from us. It does not have a fundamental strangeness. Its transcendence is wrapped in immanence."[17] Intentionality manifests a structure of immanence in the sense that the exterior world is reduced to the solipsistic ego. The illumination of objects for the subject depends on a visual economy that collapses the "interval of space" that would sustain the other's transcendence.[18] Levinas goes so far as to claim that "solipsism is neither an aberration nor a sophism; it is the very structure of reason."[19] He counters this solipsism by posing the possibility of a transcendence or exteriority that would escape the reduction of otherness to the same.

To put it all too telegraphically, light is *good* for Bennett, *bad* for Levinas and OOO. Yet must we subscribe to such a Manichean dualism between good and bad, light and dark? Consider the reading of Levinas that Derrida advances in "Violence and Metaphysics." Although he agrees with Levinas that the heliological metaphor constitutes one of the guiding lights of Western philosophy, he nevertheless interrogates his presumption that phenomenology can be utterly evaded. Derrida asks: "What language will ever escape it [the solar metaphor]? How, for example, will the metaphysics of the face as the *epiphany* of the other free itself of light? Light perhaps has no opposite; if it does, it is certainly not night. If all languages combat within it, *modifying only* the same metaphor and choosing the *best* light, Borges . . . is correct again: 'Perhaps universal history is but the history of the diverse *intonations* of several metaphors.'"[20]

OOO's stress on the withdrawal of objects into the shadowy recesses of their mutual interaction may appear to be the inverse of Bennett's solarity. Indeed, Bennett does not characterize solarity as anti-Kantian or anticorrelational, seeing "no need to choose between objects or their relations."[21] OOO's ceaseless assembling of objects at random aims to privilege objects above their relations, as if their apparent dissimilarity alone were sufficient to certify the absolute alterity of objects as they relate to each other and to us. Yet whether shadowy or luminous, objects remain phenomena for both Bennett *and* OOO. The latter wants to make *appear* the disappearance of things, to phenomenalize objects as avoiding phenomenalization. Yet this escape is untenable precisely in light of the object's withdrawal. It presupposes, in other words, the phenomenality it aims to escape precisely by illuminating the object's opacity. How might we choose the best light as Derrida suggests? Might it be something like a lunar light, an indirect luminescence that belongs neither to the violence of day nor the violence of night? Might a certain lunar judgment remain refractory to the either/ or of appearance and disappearance, presence and absence?

Solarity seems to offer a welcome antidote to the nonrelational extravagances of OOO. Yet even though Bennett wants to avoid constructing a false opposition between relations and things, she nevertheless gives scant attention to the human/object relation. Indeed, her reading of Whitman ironically betrays its own subtractive agency by bracketing the poetic "I" who licenses its ample hospitality. In order to abide the imbrication between this "I" and its others, both human and nonhuman, we must eschew the overcorrective elision of antehumanism on which Bennett's leap toward alterity depends. A more capacious reading of Whitman requires that we reckon with rather than bracket such unabashed statements as "I know perfectly well my own egotism."[22] We must assess such hubristic enunciations alongside their less boastful counterparts in order to expose a poetic voice that is equal parts intramundane and "transcendental." In "Song of Myself," for instance, the speaker asserts: "Apart from the pulling and hauling stands what I am. . . . Both in and out of the game and watching and wondering at it."[23] To be both in and out of the game is to inhabit a space of immanent transcendency in relation to the poet's litanies of countless human and nonhuman others. This immanent transcendency is synonymous neither with the sublative merging for which Lawrence chastises Whitman nor the unconditional hospitality that Bennett promotes.

Solarity's Eclipse

The distinctive cataloguing practice that pervades Whitman, his "doggedly horizontal lists," bear striking similarities to the litanies that pervade the writings of OOO theorists.[24] Ian Bogost stretches such litanizing to farcical extremes by posting an index on his blog of every reference to Mexican food from his book *Alien Phenomenology*, as well as by applying his programming skills to creating a "Latour Litanizer," a gadget that allows visitors to his website to generate at the click of a button a random list of objects derived from Wikipedia.[25] Whitman's catalogs are certainly diverse, but they are distinct from the litanizer insofar as they do not constitute an unplanned or artless "bare lists of words" (in the Emersonian phrase that allegedly inspired Whitman).[26] On the contrary, their apparent incongruity disguises an overarching organizational unity. As Bennett observes, Whitman's catalogs also *list* in the sense that they incline or lean toward various objects in a manner that is far from indifferent and unmotivated.[27] Consider the following example from "The Sleepers":

> The homeward bound and the outward bound,
> The beautiful lost swimmer, the ennuyé, the onanist, the female
>   that loves unrequited, the money-maker,
> The actor and actress, those through with their parts and those
>   waiting to commence,
> The affectionate boy, the husband and wife, the voter, the nominee
>   that is chosen and the nominee that has fail'd,
> The great already known, and the great anytime after to-day,
> The stammerer, the sick, the perfect-form-d, the homely,
> The criminal that stood in the box, the judge that sat and sen-
>   tenced him, the fluent lawyers, the jury, the audience,
> The laugher and weeper, the dancer, the midnight widow, the red
>   squaw,
> The consumptive, the erysipalite, the idiot, he that is wrong'd,
> The antipodes, and every one between this and them in the dark,
> I swear they are averaged now — one is no better than the other,
> The night and sleep have liken'd them and restored them.[28]

Employing a periodic sentence whose main clause ("I swear they are averaged now") appears after a string of subordinate ones, Whitman inscribes difference in sameness according to a pattern that Betsy Erkkila identifies with America's unofficial motto: *e pluribus unum* ("out of many one").[29] Sleep figures as a democratizing force, but its leveling effect does not precipitate the withdrawal of objects into absolute darkness. In an essay called "A Backward Glance O'er Travel'd Roads," Whitman writes that "the actual living light is always curiously from elsewhere — follows unaccountable sources, and is lunar and relative at the best."[30] The phenomenalization of objects is indirect — withdrawn as Harman would say — but not absolutely so. Luminosity is also *relational*. While "The Sleepers" employs night as an equalizing figure, "new beings appear" nevertheless as the nomadic speaker "pierce[s] the darkness," drifting among a constellation of different sleepers.[31] Their appearance is refracted through the speaker's lunar judgment that selects some objects rather than others, a judgment that does not say "yes" to everything, but necessarily discriminates.

This partiality of Whitman's catalogs, their listing or leaning toward, also manifests in the following verse from "Song of Myself," ironically perhaps given its portrayal of spontaneous human and nonhuman agency:

> The blab of the pave, tires of carts, sluff of boot-soles, talk of the
>   promenaders,
> The heavy omnibus, the driver with his interrogating thumb, the
>   clank of the shod horses on the granite floor,
> The snow-sleighs, clinking, shouted jokes, pelts of snow-balls,
> The hurrahs for popular favorites, the fury of rous'd mobs,
> The flap of the curtain'd litter, a sick man inside borne to the
>   hospital,
> The meeting of enemies, the sudden oath, the blows and fall,
> The excited crowd, the policeman with his star quickly working
>   his passage to the centre of the crowd,
> The impassive stones that receive and return so many echoes,
> What groans of over-fed or half-starv'd who fall sunstruck or in
>   fits,
> What exclamations of women taken suddenly who hurry home and
>   give birth to babes,
> What living and buried speech is always vibrating here, what howls
>   restrain'd by decorum,
> Arrests of criminals, slights, adulterous offers made, acceptances,
>   rejections with convex lips,
> I mind them or the show or resonance of them—I come and I
>   depart.[32]

Referring archaically in this context to a stretcher, the "curtain'd litter" literally conveys a sick man as well as figuratively bears the litanies that litter Whitman's verse, language that rends the curtain dividing human from nonhuman.[33] This bustling city scene of commingling human and object sounds achieves coherence not only through the "I" who surveys the scene, but also through the acoustic rhythms of the nonhuman blabs, clinks, clanks, pelts, and flaps that punctuate the verse. These sounds are collated with human hurrahs, shouts, groans, and howls: all striking examples of spontaneous, impulsive, perhaps even "bestial" speech, far removed from the humanist vision of a reasoning, intentional subject. That the "I" who "mind[s]" these things does not so much directly cognize them as feel their "resonance" stresses an aural and pulsative receptivity distinct from the appropriative mastery of visual surveillance.

Bennett plausibly reads Whitman's reference to "living and buried speech" as rendering things "vocal material actants."[34] Whitman does seem to call on us to hear things "speak" in "Song of Myself," but elsewhere objects are far less voluble. In the preface he pronounces the "beauty and dignity" of "dumb real objects," and in "Crossing Brooklyn Ferry" objects are likewise portrayed as "dumb, beautiful ministers. . . . We use you, and do not cast you aside—we plant you perma-/nently within us,/We fathom you not—we love you—there is perfection in you also."[35] That humans do not plumb the depths of objects implies a nonappropriative relation to them, yet their silent ministration to humans nevertheless appears more passive than active. In the passage Bennett discusses, moreover, the voices of objects are amplified largely on account of her deliberate silencing of human actors, who are replaced with several long strings of ellipses:

> The *blab* of the pave, tires of carts, sluff of boot-soles, . . .
> . . . the *clank* of the shod horses on the granite floor,
> The snow-sleighs, clinking, . . .
> . . . the *fury* of rous'd mobs,
> The *flap* of the curtain'd litter, . . .
> ....
> The impassive stones that receive and *return so many echoes,*
> ....
> *What living and buried speech is always vibrating here* . . .[36]

Except for the presumably human mob's "fury," Bennett quite literally encourages us to see "men and/women as dreams or dots," which is precisely how the poet in "By Blue Ontario's Shore" says the equable poet should *not* see them.[37] The reduction of humans to a typographic mark of omission conditions the humanization of things. To be fair, Bennett quotes this passage again later in the essay with most of the human actors restored. Yet she still omits the "howls restrain'd by decorum," which most nearly refers to "exclamations of women taken suddenly who hurry home and give birth to babes/What living and buried speech is always vibrating here." While Whitman frequently draws upon the trope of maternity in a manner that risks reducing the role of women to a procreative function, here the speaker disinters the pain of childbirth hidden by romanticized norms of motherhood.[38] This is not to deny the broader implications that Bennett draws from the catalogue's depiction of thingly agency, but only to question the necessity of adopting a Manichean approach whereby the illumination of objects seems to depend on the concealment of humans— in this case, women—whose literal silencing Whitman invites us to hear.

If agency is truly "*distributive*," then how are we to hear this dispersion if some of its voices are actively muted?[39]

Yet the problem here is not simply that women are silenced in favor of objects. In chapter 2 I argued that the opposition between speech and silence presumes that "giving voice" is an undeniable ethico-political good. As with the principle of solarity, this insistence on the positive value of voice is deeply Platonic insofar as it equates speech with self-presencing truth. As Andrew Cole has argued:

> [New vitalists and speculative realists] all work hard *not* to project the human into the heart of things, [but] in their attempt to respect the indifference of objects in themselves, they do so anyway by dint of the ancient Logos principle by which things call out to us and speak their being. . . . I do not see that speculative realists, or vitalists, are aware of the complicated philosophical history that underlies their project to "make things speak." Despite their attempts to question Derrida's criticism of ontotheology, this aspect of Logocentrism 101 has not been addressed.[40]

Bruno Latour's conception of nonhuman "actants," which in no small way has influenced the contemporary "object turn," is a notable example of this logocentrism. Latour writes:

> Once built, the wall of bricks does not utter a word—even though the group of workmen goes on talking and graffiti may proliferate on its surface. Once they have been filled in, the printed questionnaires remain in the archives forever unconnected with human intentions until they are made alive again by some historian. Objects, by the very nature of their connections with humans, quickly shift from being mediators to being intermediaries, counting for one or nothing, no matter how internally complicated they might be. This is why specific tricks have to be invented to *make them talk*.[41]

This desire to hear objects speak is no less guided by the principle of *language as gift* than is Susan's effort to restore Friday's voice in *Foe*, or Lestel's conception of *accréditation* by virtue of which humans who teach apes ASL are said to grant nonhumans language. Latour's *vouloir dire*, his wanting-to-say, pursues a similar auto-affective logic insofar as it echoes these demands to hear one's voice reverberated in the responsive speech of others.

As for the human agents that populate Whitman's lists, Bennett avowedly seeks to "elide or treat [them] as secondary" so that we might heed the voice of objects.[42] This strategic elision allows us "to buy time for the things outside to make their mark."[43] Solarity is thus portrayed as a temporary tactic:

> The poet must be able, periodically, to skip a beat in the regular pulse of this discriminatory perception.[44]
>
> Solar moments are necessarily fragile and fleeting.[45]
>
> The self who becomes judgment without becoming the judge suspends for a time the sociomoral categories through which s/he usually differentiates his/her responses to things.[46]

For all the references to time, however, solarity's suspension of judgment is strikingly *atemporal*. This exclusion of time is evident from Bennett's initial discussion of Bergson, whose notion of perception hinges on memory: "With the immediate and present data of our senses we mingle a thousand details out of our past experience. In most cases these memories supplant our actual perceptions, of which we then retain only a few hints, thus using them merely as 'signs' that recall to us former images."[47] Perception is subtractive for Bergson because "past images . . . constantly mingle with our perception of the present and may even take its place." He goes so far as to suggest that perception, understood as an instantaneous intuition of the external world, "is a small matter compared to what memory adds to it."[48] While Bennett briefly acknowledges the role that memory plays for Bergson, she quickly dismisses it in favor of Whitman's allegedly atemporal solarity, which "commends to us a practice of judgment unaccompanied by this image/memory."[49] To bracket memory, however, is to advance a notion of "*pure* perception" that "exists in theory rather than fact."[50] Pure perception functions as a heuristic tool for Bergson that elucidates the durational breadth of even the most rapidly occurring perception. Only the elimination of memory would permit us to "touch the reality of the object in an immediate intuition."[51] In the final analysis, solarity amounts to a mode of pure perception whose elimination of memory promises direct access to things in themselves. Solar moments are not "fragile and fleeing."[52] They are not even *moments*, that is, nonjudgmental instants wholly divorced from the past and the future. Solarity's exclusion of time recalls Husserl's dream of pure, auto-affective presence: the *augenblick* of interior monologue that excludes both time and space. Solarity seems to imagine something like an utterly pure hetero-affection, a punctual moment of total openness to alterity, divisible from the retentional past and the protentional future thanks to which perception is always partial and subtractive.[53] That solarity's "open-armed . . . impersonal" judgment depends on a nonsubtractive pure perception means that it is not "a difficult skill to master" simply because it does not qualify as a skill, which is to say a capacity for indiscriminate hospitality that humans can develop and perfect.[54] Solarity is eclipsed by time from the start.

Solar Violence

While Erkkila stresses the democratic commitments voiced by Whitman's catalogs, she also cautions that their horizontality "could operate paradoxically as a kind of formal tyranny, muting the fact of inequality, race conflict, and radical difference within a rhetorical economy of many and one."[55] As the famous "Black Lucifer" passage in "Sleepers" demonstrates, the textual insistence on sympathetic unity cannot silence the violent historical reality of slavery:

> I am a hell-name and a curse :
> Black Lucifer was not dead ;
> Or if he was I am his sorrowful, terrible heir ;
> I am the God of revolt—deathless, sorrowful, vast ; whoever
> oppresses me
> I will either destroy him or he shall release me.
> Damn him, how he does defile me !
> Hoppler of his own sons ; breeder of children and trader of them—
> Selling his daughters and the breast that fed his young.
> Informer against my brother and sister and taking pay for their blood.
> He laughed when I looked from my iron necklace after the
> steamboat that carried
> away my woman.[56]

The poet assumes the voice of a slave who rebukes the master for trading children whom he reproduces with his female slaves. Whitman removed both the reference to Lucifer's blackness and the allusion to miscegenation in the original 1855 edition of *Leaves of Grass*. In the 1881 edition he deleted the Lucifer section altogether. Scholars such as Ed Folsom and Isaac Gewirtz have detailed Whitman's ambivalent attitudes toward black Americans. Whitman was antislavery, but he did not advocate equal rights or citizenship once slavery had been abolished. In the 1850s he supported black colonization, asking "is not America for the Whites? And is it not better so?"[57] In the wake of the Civil War, he warned of "the dangers of universal suffrage," and employed familiar racist tropes in describing blacks as "but little above the beasts," possessing "as much intellect . . . as so many baboons."[58] Perhaps the most damaging contradiction to the many assertions of monistic sympathy that suffuse *Leaves of Grass* lies in an unpublished manuscript fragment, undated but most likely written late in his life: "I do not wish to say one word and will not say one word against the blacks—but the blacks can never be to me what the whites are [.] Below all political relations, even the deepest, are still deeper, personal, physiological and emotional ones, the whites are my brothers & I love them."[59] As a rejoinder to such admissions of white racial solidarity and blatant racial stereotyping, Gewirtz rewrites a line from "Crossing Brooklyn Ferry" as "I am with [some of] you."[60] In so doing, he flouts the established interpretive convention that distinguishes the "I" of a poem from its author. Whitman undoubtedly sought to impress his name and image on the reader by identifying himself as the speaker of *Leaves of Grass*, as well as by publishing the first edition with a portrait of himself in lieu of his name. The engraving of the author in laborer's clothing thus reinscribed the identity vacated by the absent signature. Is the persona named "Whitman" in the poem identical with his historical namesake, or does the former remain irreducible to the latter?

Even were one to regard Whitman's racist statements as separable from the poetic "I," however, we are still left with the paradox of a voice that declares equality by fiat. Do not such sovereign proclamations risk undermining the poetry's professed democratic ideals in a manner similar to OOO's "democracy of objects"? As I argued in chapter 3, OOO's "strong" posthumanism risks becoming the "weakest" posthumanism, which is to say the most conventionally humanist, precisely because it claims too much for itself. The clamorous claims of object independence presume the human's power to overthrow itself, thus reinstating the sovereign subject it would dethrone. Yet Whitman's self-conscious performance of sovereign power lends itself to an ironic reading that contrasts sharply with object-oriented ontology. After all, Whitman's "I" *knows* its own egotism. Should we take at face value a poet who declares "I will effuse egotism and show it underlying all, and I will be the bard of personality"? Is it precisely this ironic element that Lawrence missed when he chided Whitman's arrogance? Far from "preaching egotism," Whitman explained to his friend Horace Traubel that the "I" of his poetry designates:

> personal force: it is personal force that I respect—that I look for. It
> may be conceit, vanity, egotism—but it is also personal force. . . . It is
> of the first necessity in my life that this personal prowess should be

brought prominently forward—should be thrown unreservedly into our work. If I said "I, Walt Whitman" in my poems and the text meant only what it literally said, then the situation would be sad indeed—would be very serious: but the Walt Whitman who belongs in the Leaves is not a circumscribed Walt Whitman but just as well a Horace Traubel as any one else—personalized moral, spiritual, force of whatever kind, for whatever day; it is force, force, personal force, we are after.[61]

Despite the sixfold repetition of the word *personal*, this force is remarkably *depersonalized*, inhabitable by anyone who assumes the position of the poetic "I," which means that this position is precisely nowhere in particular, personal only by virtue of the I's contingent self-enunciation, and thus radically impersonal and anonymous.

The force of this (im)personality expresses what Derrida characterized in his early essay, "Structure Sign and Play in the Discourse of the Human Sciences," as the "contradictorily coherent" identification of a center that "governs the structure" but does not fully belong to it: "Coherence in contradiction expresses the force of a desire."[62] The metaphysical disavowal of the center's structurality is identical to the logic of sovereignty that Derrida would theorize many years later in *Rogues*. The sovereign governs from the center of a structure to which he does not belong, totally inside/outside, totally "alone" and exceptional. In order to maintain this contradictory position, pure sovereignty must remain utterly silent. As soon as this silence is broken, as soon as sovereignty gives reasons to justify its exceptionality, sovereignty becomes shareable. Its "center" is elsewhere. Sovereignty is no longer sovereignty. Yet the dispersal of the "I" in Whitman is distinct from metaphysical sovereignty insofar as Whitman is far from silent about it. He is all too ready to declare its performative contingency: "Do I contradict myself?/Very well then I contradict myself,/(I am large, I contain multitudes.)."[63] The contradictory coherence of Whitman's declarations goes to the fundamental antinomy that Derrida identifies at the center of democracy itself: the aporetic knot formed by a freedom that secures the sovereign self at the "center" of a political structure of equality shared by many other selves. Drawing from Aristotle's canonical theory of politics, Derrida underscores how democracy wavers between an infinite, incalculable freedom on the one hand, and a delimited, calculable equality on the other.[64] Absolute, unfettered freedom would be antidemocratic insofar as it would disallow the political conditions that safeguard freedom's equal distribution. Every citizen is a potential rogue whose otherwise unbridled sovereignty is restrained by the principle of equality. Rather than abide the unconditional sovereignty of each citizen, democracy requires, at least in principle, that everyone be equally free, thus circumscribing liberty from the start.

Whitman's democratic desire might seem more in line with the mock Latin *e unibus pluram* (out of one, many) than *e pluribus unum*.[65] Yet sovereignty and democracy do not find themselves suspended *between* mutually exclusive demands: an incalculable freedom facing off against a calculable equality. Rather, they are suspended within, by, and from themselves. Sovereignty is internal to democracy, and vice versa. Insofar as everyone is equally free, equality is never purely calculable and freedom is never purely incalculable. Derrida draws from medical vocabulary to account for how democracy is afflicted with an "autoimmune" conflict between unconditional freedom and conditional equality, a coupling by virtue of which freedom and equality always bequeath both conditional and unconditional traits in whatever concrete forms they assume.[66] This internal division by virtue of which the many is always one and the one always many accounts for why Whitman expresses his democratic desire as a "personal force," the *kratos* (*-cracy*, "force") of a singular subject whose power is nevertheless dispersed among numerous others.

Whitman's poetic voice often seeks to expand the boundaries of democracy beyond the "we" that tacitly defines itself as white and male, most notoriously in the Declaration of Independence and the United States Constitution. Yet an expanded democracy is not and cannot ever be wholly inclusive. The question of *who* or *what* belongs to democracy is infinitely reassessable because there are no natural limits that determine its borders once and for all. Democracy is founded on both explicit and implicit discriminatory judgments regarding who or what are to be included. Derrida remarks, for example, that the extension of voting rights to immigrants is not necessarily more democratic: "One will never actually be able to 'prove' that there is more democracy in granting or in refusing the right to vote to immigrants, notably those who live and work in the national territory. . . . One electoral law is thus always at the same time more and less democratic than another."[67] Later he asks, "how far is democracy to be extended, the *people* of democracy, and the 'each "one"' of democracy? To the dead, to animals, to trees and rocks?"[68] Despite appearances, Derrida is not proclaiming the advent of a cosmocracy that would include everything; rather, the interrogative form of this ostensibly inclusive gesture stresses that the problem of democracy's limit must *remain* a question. We might *believe* that we are being "more democratic" when we include trees and rocks, but this irreducible belief cannot be converted into a definitive calculation that would evidence our perfectible march toward absolute inclusivity. What about bacteria and viruses? Artificial intelligence? The point is not that we should include everything, but that the inclusion of whomever or whatever cannot be construed as "more democratic" except by conceiving democracy as a totality—except, that is, by conceiving democracy *undemocratically*. The calculation of "more" and "less" corresponds to a container model of democracy. To speak of political measures and decisions as more or less democratic implies an all-inclusive ideal against which this or that other can be added or subtracted.

The impossible advent of full inclusivity leads Derrida to conclude that democracy always remains *to come*. This deferred democracy does not correspond to a regulative idea: an aspirational yet unachievable state of perpetual peace and inclusivity toward which we nevertheless ought to aim. Nor does this deferral constitute an alibi, a "right to defer." On the contrary: "[T]he to-come of democracy is also, although without presence, the hic et nunc of urgency, of the injunction as absolute urgency. Even when democracy makes one wait or makes one wait for it."[69] We are ethically obliged to promote an imperfect democracy even though it never fully presents itself.

Whitman's incessant poetic listing undeniably *lists* (desires) democracy. It yearns for a shared, all-inclusive world that awaits only the elimination of sociopolitical hierarchies. Although his litanies are not composed of random, unrelated objects, they nevertheless share with OOO a teleological desire to level all hierarchies and antagonisms. In contrast with Derrida's democracy to come, Whitman's democracy "arrives" by force of his sovereign enunciations. In "Thou Mother with Thy Equal Brood," for instance, America is figured as the "ship of democracy" whose freight bears "Earth's *résumé* entire."[70] America is the "living present brain, heir of the dead, the Old/world brain," a world of "superber birth" nurtured in the womb by Europe and Asia, those "antecedent nations" that beget America as its finest offspring.[71] The speaker concedes that the ship is "not to fairsail unintermitted always, / The storm shall dash thy face, the murk of war and worse than / war shall cover thee all over."[72] In addition to foreseeing the rough seas ahead, the speaker also acknowledges that one can only limn the future of American democracy:

> Thou wonder world yet undefined, unform'd, neither do I define
> thee,
> How can I pierce the impenetrable blank of the future?

I feel thy ominous greatness evil as well as good. . . .

....

But I do not undertake to define thee, hardly to comprehend thee,
I but thee name, thee prophesy, as now,
I merely thee ejaculate![73]

Whitman fittingly couples the insemination of American democracy with metaphors of containment. From the nation's "teeming womb" emerge "giant babes [the states] in ceaseless procession." America is "the globe of globes," an "orb" that "the FUTURE only holds."[74] This spherical conception of democracy figures America as the living "heir of the dead, the Old/World brain," which incubates the new world until its birth.[75] Crucially, this matrix is inverted once America is born: nourished in the womb of the old world, the new world henceforth becomes "the globe of globes" that fully incorporates the past into its present, ideal form.[76]

Although the poem affirms the vulnerability of democratic progress, this problem is quickly cast overboard. Come what misfortunes may, America will "surmount them all."[77] Citing America's artistic, educational, spiritual, and moral superiority, the speaker declares, "These! these in thee, (certain to come,) to-day I prophesy."[78] In contrast with the disseminated, unguaranteed future of the democracy to come, the content of Whitman's future America is predetermined: the poet prematurely "ejaculates" a democracy "certain to come."

The "ship of democracy" thus names a vessel whose unwavering desire is vulnerable only to external forces, figured as a "livid cancer" that threatens with its "hideous claws, clinging upon thy breasts, seeking to strike thee deep within."[79] We might be tempted to read this malignancy in terms of Whitman's exceptionalist ideology, which encloses democracy within a "global" America that excludes the rest of the world. America is said to eclipse all other nations because it is the unique "fruit of all the Old ripening."[80] Alternatively, we might attribute this cancer to Whitman's racism or the sociopolitical inequalities of nineteenth-century America more generally. Yet to focus only on these threats from without is to disavow the autoimmunitary struggle between freedom and equality internal to democracy as such. Infinitely perfectible and infinitely corruptible, democracy is deferred not simply due to accidental, "contaminating" forces such as American exceptionalism or racism. Even were Whitman to have cast a wider net and have recognized the contributions of other nations to his democratic ideal, even were he not to have expressed racist views, his conception of democracy would remain no less teleological, no less committed to a democratic desire that knows itself, a desire seemingly undivided in its aim to transcend inequality altogether. As Martin Hägglund shows in his analysis of the Derridean *à venir*, the arrival of absolute justice or a perfect democracy would annul the temporal flux on which the survival of democracy depends. Democracy must make itself available to change and transformation. It would cease being democratic as soon as it completely resolved its internal conflicts. The full and final instantiation of democracy is undesirable because it would ultimately threaten democracy precisely by impeding the mutability and variability on which it thrives. The "perfect" democracy would be the worst democracy: a democracy that negates itself by insulating itself.

The Cosmocracy to Come

The perfect democracy may be undesirable, yet the impulse to achieve its realization remains no less robust, especially in light of what I described in the introduction as the whataboutism of contemporary posthumanist thought. Given this ampersand effect, this ceaseless inclusion of more and more to which contemporary theory commits itself, does it still make sense to talk about democracy if this word no longer limits itself to human *dēmos*? In *Plant Thinking*, for instance, Michael Marder advocates a "vegetal democracy" that would be "open not only to *Homo sapiens* but to all species without exception."[81] Not content to stop at plants, the theoretical turn toward inanimate things wants to navigate the compass even farther. The title of an essay by Timothy Morton sums it up: "Here Comes Everything: The Promise of Object-Oriented Ontology."[82] This promise is similar to Marder's "phytocentrism to come," which draws upon the decentered nature of plants (their lack of a central nervous system) in order to expand the sphere of vitality to include all life. Phytocentrism's paradoxical acentrism stands as a synecdoche for the growth capacity shared by every living species. As with the proclamation "here comes everything," which does not simply describe but also performatively solicits the advent of its predicate, phytocentrism obeys a conventional messianic logic by explicitly envisioning its success.[83] The Derridean *à venir*, by contrast, garners its strength precisely from its weakness; its lack of assurance makes its promise more robust than any declaration of infinite hospitality insofar as deconstruction says "yes" to a future that remains vulnerable to both chance and threat.[84]

That Derrida ponders what it might mean to extend democracy to the dead as well as the living, the inanimate as well as the animate, underscores that the *à venir* already invites the cosmocracy to come insofar as the *kratos* is severed from its essential relation to the *dēmos*. As he makes clear in an interview with Michael Sprinker, Derrida is not exclusively attached to the term *democracy*. He accepts its legacy on account of its exposure to the future, its promise of an open-ended negotiation between equality and freedom, but he also acknowledges:

> Perhaps the term democracy is not a good term. For now it's the best term I've found. But, for example, one day I gave a lecture at Johns Hopkins on these things and a student said to me, "What you call democracy is what Hannah Arendt calls republic in order to place it in opposition to democracy." Why not? I am only employing the term democracy in a sentence or a discourse that determines certain things. I think that in the discursive context that dominates politics today, the choice of the term that appears in the majority of sentences in this discourse is a good choice—it's the least lousy possible. As a term, however, it's not sacred. I can, some day or another, say, "No, it's not the right term. The situation allows or demands that we use another term in other sentences." For now, it's the best term for me. And choosing this term is obviously a political choice. It's a political action.[85]

If accepting the paleonym democracy in order to renew its promise constitutes a political choice, then adopting the neopaleonym *cosmocracy* is no less politically strategic. Is this term less "lousy" than *democracy*, which limits its scope to presumably human *dēmos*? When Derrida contemplates a potentially inhuman political force, he does not specifically advocate on behalf of the insentient and the inanimate. He opens the door to rocks and trees, in other words, but it seems only slightly ajar. Can we speak of petrological life? This query does not require an immediate, urgent response so much as it welcomes a porous cosmocracy to come that eschews democracy's petrified forms.

Distinct from unconditional hospitality, cosmocracy renews the promise of an infinitely expansive inclusivity by neither declaring nor ensuring its achievement. Its porosity is also its aporosity. That unconditional hospitality is a principal rather than a practice, however, does not mean that it can be dismissed altogether. "Unconditional hospitality is impossible," as Michael Naas puts it, but "it is the only hospitality that can give any meaning to the concept of hospitality itself and, thus, the only possible hospitality, the only one *worthy of this name*."[86] On the other hand, unconditional hospitality can never quite live up to its name because its practice is circumscribed by a negotiation that takes place between a finite number of hosts and guests. Geoffrey Bennington frames this aporia as follows: "The one hospitality (the unconditional one) is worthy of the name because it just is itself, coincides with what the name names; the other (the conditional and conditioned) is worthy of the name in the sense that it is done *in the name* of the other, unconditional hospitality. . . . An act of hospitality is worthy of the name not because it simply coincides with its name but because it is done *in the name* of what it never quite is."[87] Bennington notes that this sense of falling short is very close to the Kantian regulative idea even though it cannot ultimately be reduced to it. How are we to keep alive the promise of the cosmocracy to come if not by striving toward some infinitely receding horizon? If this promise is not teleological, then it cannot, by the same token, be absolutely *anti*teleological. Every promise presupposes some aim, even if its content remains devoid of univocal meaning.

That the *à venir* is neither teleological nor antiteleological seems confirmed by the almost confessional tone that Derrida briefly adopts in *Rogues*: "The regulative Idea remains, for lack of anything better, if we can say 'lack of anything better' with regard to a regulative Idea, a last resort. Although such a last resort or final recourse risks becoming an alibi, it retains a certain dignity. I cannot swear that I will not one day give in to it."[88] Even though the *à venir* is not a principle that guides us toward some impossible ideal, Derrida cannot reassure us that he will not get buoyed away by its aspirational drift. Much depends on how one reads "Je ne jurerais pas de ne jamais y céder."[89] Michael Naas and Pascale-Anne Brault translate this sentence as "I cannot swear that I will not one day give in to it." Geoffrey Bennington renders it more literally as "I would not swear never to give in to it," a translation that has the benefit of retaining the temporal ambiguity of the original French.[90] Has Derrida already yielded to the regulative idea? Does this describe not so much a resignation that could happen *one day* (in the future) as one that has already happened? Bennington also offers an alternative rendering whose temporality is even more difficult to pin down: "I would not swear that I never give in to it": in general, on an ongoing basis, I cannot pledge that I have not been tempted to surrender to teleological yearnings.[91]

Lest any reader jump too quickly to the conclusion that this sentence amounts to a "gotcha moment" that utterly undermines Derrida's efforts to separate the *à venir* from the regulative idea, it is crucial to reiterate that the *à venir* does not set itself in opposition to the Kantian idea even as it remains irreducible to it. The *à venir* is against the Kantian idea in a double sense: the former works against and along with the latter, touches (on) it, sympathetically perhaps, even as it comes up against and thwarts its aims.

If Derrida admits to falling short of rejecting the Kantian idea—which amounts to saying that he falls short of not falling short—this miscarriage needs to be understood as something other than failure in the usual sense: what does it mean *to fail* to fail, to miss by not missing, or rather, by *nearly* missing the target toward which one all along claimed not to be aiming? Such vertiginous formulations aim to reconceive *falling short* as something other than a privation. Far from conceding a seduction by teleology that has already occurred or might occur at some point in the future, Derrida expresses nothing less than the autoimmunitary betrayal by virtue of which the rejection of teleology is no less vulnerable to contradiction than its affirmation.

This contradiction is legible on the very surface of the expressions *democracy to come* or *cosmocracy to come*, both of which depend on the principle of "minimal linguistic transparency" discussed in chapter 1.[92] Recall that for Derrida the intelligibility of any signifier depends on a reduction to univocity that "must be recommenced indefinitely, for language neither can nor should be maintained under the protection of univocity."[93] This reduction to univocity must be forever resumed because irreducible equivocity is the condition of possibility and impossibility for univocity, and vice versa. The *doit* of Derrida's reduction to univocity is both quasi-transcendental and normative: the reduction must and *should* be recommenced indefinitely because univocity cannot and *should not* protect itself from equivocity.[94] As Leonard Lawlor puts it, "there is an irreducible inadequation between possibility and necessity."[95] The imperative of univocity attempts to put the brakes on an unstoppable equivocity. As Derrida observes, even Joyce's radically equivocal prose depends on a minimal univocity whose failure demands its interminable renewal.

As an iterable term, democracy summons up a history of sedimented meanings whose intelligibility depends on it being at once absolutely translatable and absolutely untranslatable, totally univocal and totally equivocal. This oscillation is the necessary condition of its promise. Democracy may lack a "proper form," but it must and should be promised as a minimally transparent, iterable concept in order that it have any future at all.[96] On the one hand, democracy is fundamentally "aporetic in its structure (force without force, incalculable singularity and calculable equality . . . indivisible sovereignty and divisible or shared sovereignty, an empty name, a despairing messianicity or a messianicity in despair." On the other hand, one can despair of this messianicity, one can lament the hopeless advent of the *à venir*, only on account of having first supplied this "empty name" with some intelligible meaning, some provisional sense that permits us to reactivate the names democracy, equality, hospitality and so on. This reactivation would differ from its Husserlian understanding, which Derrida glosses as reawakening a "primordial sense" that has been buried underneath sedimented tradition.[97] Derrida *reactivates* reactivation in a manner that abides the accumulated historical residue whose effacement Husserl holds onto as a horizon of impossibility: a guiding, aspirational ideal of unattainable univocity.

Wanting Justice

What does this simultaneous translatability and untranslatability reveal about the *desire* for the cosmocracy to come, this "empty name" that nevertheless harbors some promise, one that neither yields to the Kantian idea nor frontally opposes it? To address this question, I want to revisit the claim advanced by Hägglund that the struggle for a more just and inclusive world is not governed by a desire for absolute plenitude. According to Hägglund:

> Every call for justice must affirm the coming of time, which opens the chance of justice and the threat of injustice in the same stroke. The desire for justice has thus never been a desire for absolute justice. The desire for justice is always a desire for the survival of finite singularities, which violates the survival of other finite singularities. Every ideal of justice is therefore inscribed in what Derrida calls an "economy of violence." To be sure, struggles for justice are often perpetrated in the name of absolute justice, but these claims can always be shown to be incoherent and hypocritical. There is no call for justice that does not call for the exclusion of others which means that every call for justice can be challenged and criticized. The point of this argument is not to discredit calls for justice but to recognize that these calls are always already inscribed in an economy of violence.[98]

Hägglund wants to provide a "systematic account" of desire in Derrida that neither he nor his commentators have explored, a unifying logic that allegedly subtends his conception of the *à venir*.[99] In developing this account, Hägglund employs a conception of undesirability that oscillates between description and prescription. Absolute justice is undesirable in a prescriptive or normative sense because it would result in an unwanted outcome: the erasure of the antinomy between freedom and equality whose survival depends on the chance of perfectibility and the threat of corruptibility. If democracy is sustained by an irremediable autoimmunitary conflict between freedom and equality, then the "perfect" democracy would require the complete erasure of alterity: a Robinsonian isolation whereby "equality" would be fully reconciled with an unconditional, sovereign freedom. Although Derrida does not explicitly state that the perfect democracy is undesirable in a normative sense, the deleterious effects that would follow from the arrival of a democracy "cured" of its autoimmunity support the claim that the twofold chance and threat to which we must remain open implies a distinction between the desirable and the undesirable.

Shifting to a descriptive level, however, Hägglund wants to claim that absolute justice is undesire-able in the sense that we are *unable* to desire it: "The desire for fullness *has never been operative* in a political struggle or anything else."[100] The age-old admonishment to be careful what you wish for is utterly gratuitous because Hägglund believes no one really desires political plenitude. He is certainly correct that the fight for pure justice "can always be shown to be incoherent and hypocritical," but why can we not desire absolute justice simply because this yearning is self-refuting?[101] It does not follow from the principle of democracy's intrinsic mutability that "one cannot desire a state of being that is exempt from time" simply because this desire is contradictory.[102] Since when is desire *not* divided, incoherent, and contradictory?[103] One of the most basic lessons of psychoanalytic theory is that desire is forever mobile and shifting, never entirely certain of its object. That Hägglund is not unaware of this fundamental psychoanalytic insight is made clear when he claims that the desire for democracy "is essentially corruptible and inherently violent."[104] This corruptibility of desire, however, proves the exact opposite of what he presumes. Far from unveiling the true sense of the desire for justice, far from exposing its "proper" object, the corruptibility of desire renders untenable any desire to distinguish real from false desires. Notwithstanding assertions to the contrary, this false distinction between true and false desires cannot but "discredit calls for justice" because it refuses to *credit* the desire for perfect justice as a desire at all.[105] Hägglund thus seizes upon Derrida's association of democracy with mutability and openness, susceptibility and vulnerability, but this commandeering of democracy's ship ironically protects it from straying off course. Despite claiming to promote a "hyper-political logic that spells out that nothing is unscathed or unquestionable," he exempts desire from corruptibility and autoimmunity, which is to say that he exempts it from temporality altogether.[106]

One can only translate the untranslatable *sense* of the desire for democracy into an absolute transparency by force of a sovereign desire to unveil the truth of desire: "To desire democracy is *by definition* to desire something temporal, since democracy must remain open to its own alteration in order to be democratic [my emphasis]."[107] This effort to pin down the meaning of the desire for democracy is far removed from the minimal transparency or univocity on whose necessity Derrida insists. Although the intelligibility of the desire for democracy depends on it "allotting its share to univocity," this minimal transparency neither can nor should be protected from desire's irreducible equivocity. Desire is by definition indefinable. The aim of desire, its proper object, is always *à venir*. To identify its aim here and now is to grant language a referential function fundamentally at odds with *différance*. The desire for desire's referential truth disavows the iterability of the sign, its structurally intrinsic alterity that depends precisely on the coming of time whose twofold chance and threat Hägglund otherwise wants to affirm.[108]

The delimitation of the desire for democracy is undemocratic not simply because a sovereign agent appoints himself the arbiter of true and false desires, but more fundamentally because this "proper" definition disavows desire's improperness, its inherent dispossession. That desire is "the desire of the other," as Kojève and Lacan have taught us, means that it never fully belongs to us.[109] Mediated through the desires of others, "our" desire is fundamentally expropriated. Desire is never sovereign for the same reason that sovereignty is never sovereign: both are always already divided by alterity. No one indeed has ever desired absolute justice, but not because the predicate of this desire is fully determinable. No one has ever desired absolute justice—or desired anything else for that matter—because desire has always belonged to no *one*.

The drive to unveil the alleged descriptive undesirability of unconditional justice thus overlooks a far more radical undesirability. When Lacan and Kojève claim that we *desire* desire, they figure desire as the object of a desiring subject. This capacity for desire is said to elevate humans above animals, whose aims are allegedly reducible to self-preservation. While Kojève employs the term *desire* to characterize both human and animal intentionality, Lacan less charitably distinguishes human desire from animal "need."[110] Animals lack desire just as they lack language. Yet desire is not merely one object of desire among others. Humans no more "have" desire than they do language. Lack is the condition of desire, but desire also names what we lack. Desire is a privation rather than a possession. Desire is an *impouvoir*, a nonpower or not-being-able.[111] This *impouvoir* is not the same as powerlessness. It corresponds to a certain weakness and vulnerability, an exposure to forces beyond ourselves to which we yield, for better or worse.[112] As Derek Attridge puts it, desire is an *arrivant*: it comes

from elsewhere, from the alterity of conscious and unconscious investments whose ownness is permanently displaced.[113] Desire is not an ability
or capacity that belongs to an agential subject. An originary undesirability
is the condition of impossibility of desire as such. *Desire is undesirable.*

The Scandal of Desire

The cosmocracy *to come* is both *hic et nunc* and infinitely deferred. Promised in the name of what never fully arrives, cosmocracy is worthy of the
name only insofar as it remains unworthy of the name. Whereas the antiteleological desire for a continuously perfectible and corruptible justice to
come confidently asserts its incorruptibility, its capacity not to surrender
to any teleological yearnings, the weak nonteleological desire for the cosmocracy to come is far less self-assured and confident, far less eager to
pledge its unswerving desire never to fall short of falling short.

Whitman's desire to broaden democracy beyond the human warrants
redescribing this impulse as cosmocratic. While this desire is just as resil-
ient as Hägglund's antiteleological counterpart, occasionally Whitman
loses his swagger and concedes cosmocracy's fragility and vulnerability.
Leaves of Grass employs a central botanical metaphor of singular leaves
composing a common, democratic ground. Yet this harmony is not always
welcomed by the poetic voice. In "This Compost," for instance, the speaker
withdraws from the earth because he is repulsed by the thought that corpses
are buried within it:

> O how can it be that the ground itself does not sicken?
> How can you be alive you growths of spring?
> How can you furnish health you blood of herbs, roots, orchards,
>   grain?
> Are they not continually putting distemper'd corpses within you?
> Is not every continent work'd over and over with sour dead?[114]

Here the speaker eschews earthly union in a manner that ironically recalls
Lawrence's repudiation of Whitman: "I don't want all those things inside
me, thank you." Whitman seems to reject the "foul liquid and meat" of
the dead just as Lawrence does the "awful pudding of One Identity."[115]
While critics following Lawrence's lead have often admonished Whitman for promoting an "'imperial self,' an ego spreading outward,"
Killingsworth suggests that a different Whitman appears in these verses,
"a poet of limits."[116] How are we to understand these borders? Are they

reducible to the xenophobia that we have seen elsewhere in Whitman? To be sure, something of the cosmocratic promise remains once the speaker recognizes:

> Perhaps every mite has once form'd part of a sick person—yet
>   behold!
> The grass of spring covers the prairies,
> The bean bursts noiselessly through the mould in the garden,
> The delicate spear of the onion pierces upward,
> The apple-buds cluster together on the apple-branches,
> The resurrection of the wheat appears with pale visage out of its
>   graves.[117]

These lines seem to revitalize botanical merger, yet the speaker never entirely recovers from his initial xenophobia: "Now I am terrified at the Earth, it is that calm and patient, / It grows such sweet things out of such corruptions."[118] The speaker's expressed fear of "every spear of grass"— which is to say every singular other who/that threatens to pierce "the imperial self"—can also be read as conveying a more literal terror in the face of what Killingsworth calls earth's "thingishness."[119] Insofar as the earth continues to rotate with or without us, its inhuman indifference is an unwelcome reminder of human mortality. The earth "turns harmless and stainless on its axis, with such endless/successions of diseas'd corpses, / It distills such exquisite winds out of such infused fetor, / It renews with such unwitting looks its prodigal, annual, sumptu-/ous crops, / It gives such divine materials to men, and accepts such leavings/from them at last."[120] Barring a collision with a rogue planet such as Melancholia, the earth will survive the death of each finite singularity, at least according to the poet's vision of the world's "stainless," eternal rotation.[121]

Yet perhaps the earth also terrifies the poet precisely because its eternal survival, its continuous turn, is no more guaranteed than that of democracy, which, as Derrida frequently insists, is defined by:

> [a] rotary motion . . . that turns on itself around a fixed axis. . . . It
> seems difficult to think the desire for or the naming of any democratic
> space without what is called in Latin a *rota* . . . without the rotary
> motion of some quasi-circular return or rotation toward the self,
> toward the origin itself, toward and upon the self of the origin, when-
> ever it is a question, for example, of sovereign self-determination, of
> the autonomy of the self, of the *ipse*, namely, of the one-self that gives
> itself its own law.[122]

In addition to this turn toward the self as origin, democracy's circularity also turns on a lack of proper meaning that operates like "a disengaged clutch, freewheeling. . . . And so it is defined only by turns, by tropes, by tropism."[123] Moreover, this tropism is likewise inscribed in the turn *qua* alternation between freedom and equality, an oscillation that does not simply mark a torsion between the unconditional and the conditional, but also the turning of these concepts within and against themselves.

Whitman similarly figures cosmocracy in terms of the irresistible revolution of the earth, a desire whose "true" meaning cannot be arrested: a perpetual tropism that defies any Ptolemaic effort to center the meaning of desire, to prevent the desire for absolute justice from turning on or against itself. One can thus draw an axis directly from the desire to eliminate injustice all the way through to the sovereign self-determination to eliminate the desire to eliminate all injustice. To claim that a perfectly just world is undesirable because it would result in an atemporal, inalterable state is to say that we cannot want it because it will terminate the rotational movement on which democracy, as it were, rests. Absolute justice is thus retained as a regulative idea, only now its purpose is overturned: its portended *injustice* is meant to regulate us into not wanting it precisely because it has been found wanting. Yet it can only be *found* wanting by claiming to have unearthed its true meaning at the expense of leaving unturned its other tropic leavings.

One may pledge never to yield to teleology, but this desire not to be ensnared is no less corruptible than the most explicit affirmation of absolute justice itself. A promise that protects itself from corruptibility is no promise at all. Corruptibility has no opposite: no incorruptible side that would reveal the truth of the desire that animates every promise. The "perfect" egalitarian ethico-political condition may amount to a solipsistic nightmare in which freedom and equality coincide only on account of eliminating all others. That no desire is utterly transparent to itself, however, means that we cannot completely rule out the possibility that this seemingly undesirable state nevertheless names one possible desire among others.

Although we will never be in a position to know that we have included everything, it does not follow that we cannot want the impossible. It no more falls to us to determine the truth of our desires then it would to the "sovereign" human to adjudicate cosmocracy's achievement. This judgment would ironically close things off by granting the human the power to calculate the final sum of those deemed worthy of ethical consideration. The necessary porosity of cosmocratic borders resonates to some extent with Matthew Calarco's notion of an "agnostic ethics," a Levinasian inspired openness to the question of who or what has a face.[124] This agnosticism is not based on "the positive claim that all things or all life forms do count."[125] On the contrary, he maintains that the question of ethical consideration ought to remain "wide open."[126] Calarco's conception of openness, however, tends to mute the role of decision and calculation in ethics: "We are obliged to proceed from the possibility that anything might take on a face. And we are further obliged to hold this possibility permanently open" because the scope of ethical concern "cannot be determined with any finality."[127] The borders that demarcate the sphere of ethical concern may not be fully decidable, but this indetermination cannot and should not warrant avoiding decisions or refusing to invoke specific criteria as a condition of moral consideration—however contestable and revisable such criteria always remain. We must "decide" the undecidable, which means that ethical agnosticism is contaminated from the start. Calarco seems to concede as much when he remarks that his decision to focus on animals risks undermining universal ethical respect. He is right to resist the tendency of analytic philosophers to invoke criteria that "cleanly demarcate those beings who belong to the community of moral patients from those beings who do not [my emphasis]."[128] Yet this lack of clean demarcation cannot be the basis for abandoning criteria altogether. The criterion of sentience, for example, tacitly subtends Calarco's attention to animal suffering. The inescapable impurity of ethical agnosticism entails that we are "bound to make mistakes" when we exclude some beings from moral consideration, even though Calarco explicitly suggests that agnosticism can assist us in avoiding them.[129] Only an impossible, unconditional agnosticism could avoid mistakes altogether. We must strive to preserve the possibility that anything might have a face, even though we continually "decide" that some beings and entities do not according to a range of criteria that calculate the incalculable. Have not both omnivores and vegetarians already said "no" to plants by eating them? Do I know that plants do not have a face? No. But I am reasonably confident in my belief that they do not, certain enough that my (irreducible) belief rightly passes for knowledge.

The question of belief returns us to what I characterized in chapter 1 as the co-insinuation of the *as if* and the *as such*—the phantasm and its corresponding "truth." That mistakes are inevitable means that this book has proceeded *as if* its chosen literary texts and films are more germane to the question of the human's place within posthumanism than others might have been. In contrast to the cold calculations of the litanizer, my textual selections reflect a leap of faith whose lack of assurances affirms discriminatory, "arbitrary" decisions as the condition of hospitality as such. As I noted previously, this arbitrariness should not be equated with pure chance, but rather with judgments that arbitrate according to beliefs whose veracity remains continuously debatable. The weak posthumanism of these decisions does not hesitate with regard to the human's urgent decentering. It does not convey a desire (at least a conscious one) to safeguard the human from its deconstruction. On the contrary, its weakness reflects an ironic power that aims to deflate the ultrahumanism of those posthumans who "know" too much about the human. They are ultimately too little of faith precisely because they credit the human's capacity for self-abandonment as if it amounted to the truth of posthuman capaciousness as such.

Capaciousness *is* capability itself. There is no making room for the non-human without a human giving itself the capacity to play host at the "cosmic party" to which everything is allegedly now invited.[130] I began by considering how posthumanist theories of immanence circumnavigate the human only to run aground on their scandalous desire to evade the scandal of the human. Are we now any further from the shores of the human than when we embarked? What will have been the measure of our success or failure? Has the human been erased "like a face drawn in the sand at the edge of the sea"?[131] Or does its successful effacement remain no more determinable than the origin of the footprint whose trace disturbs Crusoe's solitude? Surely we hope that we have landed on the archipelago of a lesser or weaker humanism even if we remain linked to the island of human narcissism from whence we came.

Posthumanist immanence boards the ship of cosmocracy hoping to sail unhindered through the aporia of the human, the nonpassage that in Greek is often associated with inaccessible waterways and seas.[132] Yet the desire to steer around the desire for an entirely equitable, nonhierarchical world is no less encumbered. It seeks to make land according to its "logic" of desire, which is to say its logos of desire, precisely by making fast desire's irrevocably unmoorable aims. We must therefore not conclude that no one has ever heard the posthumanist siren song, that its seduction amounts to little more than a false consciousness that awaits us to fathom its true, submerged desire. While some posthumanist theories demand the imminent arrival of an *immanent*, untroubled ethico-political condition, others posit it as belatedly achievable once the human is fully decentered. This deferred dream thereby seeks to make the future present by "knowing" its destination in advance. It may seem that this long-sought perfect justice lies leagues apart from the desire to repudiate this yearning altogether, yet both desires are no more impervious to time's roll and pitch than is the promise of justice itself. Whereas the ship of democracy requires a helmsman with a "good strong hand and wary eye" to steer it toward its foreseeable and inevitable landing, the ship of cosmocracy sails toward justice in a manner that cannot protect it from the elements.[133] This ship can never perfectly ballast the competing demands that issue from countless others—human or animal, organic or inorganic—beings and entities who call on us to see the cosmocratic vista anew. They implore us not only to reexamine who or what we ought to embrace, but also to limn the fading horizon that separates whom from what. Navigating the rough waters on which we both steady and list, we must tack our ship directly into the erratic winds by whose force justice sinks or swims—ceaselessly borne back toward the cosmocracy always to come.

ACKNOWLEDGMENTS

A number of individuals supported the research and writing of this book. In addition to reading sections of the manuscript, Chris Eagle conceived the title. I resisted it initially, but eventually came around to embrace it. I am grateful to the following friends and colleagues who donated their time to read drafts at various stages of the project: Chris Fleming, Lindsay Tuggle, Diego Bubbio, Chris Danta, Peggy Kamuf, Bonnie Honig, Margaret Harris, and Lori Marso. Their thoughtful responses were invaluable in strengthening the book's argument. I am always especially thankful for eleventh-hour interventions, and on this occasion I owe Jennifer Milne for drawing my attention to the gendering of von Max's painting (not to mention the many delicious dinners provided by her and James Morley). Several chapters were first presented at Brown University, DePaul University, and the University of New South Wales. On each occasion my work was greeted with an enthusiasm and keen discernment that challenged and honed my thinking. As always, I thank my family: Linda Hoffman-Peterson, John Peterson, Ann Peterson-Miller, Elizabeth Peterson, and Geoffrey Peterson.

I am especially fortunate to have been guided by the attentive editorial direction of Thomas Lay at Fordham University Press. I am also indebted to J. Hillis Miller and two anonymous reviewers for reading the manuscript and providing insightful suggestions for revision. The sudden loss of Helen Tartar in 2014 was shared and mourned by many Fordham authors. I find some consolation in believing that she would be pleased with how Tom and Richard Morrison have ably carried forth her extraordinary legacy. I also want to thank Michael Koch for his vigilant editorial support during the book's production and Suzanne Gupps for her assistance with the index.

Chapters 2 and 3, as well as a section of chapter 1, were previously published in modified form. A shorter version of chapter 2 was published under the title "The Home of Friday: Coetzee's *Foe*," in *Textual Practice*, online first, 2015; reprinted by permission of the publisher Taylor and Francis Ltd., www.tandfonline.com. A shorter version of chapter 3 was published as "The Gravity of Melancholia: A Critique of Speculative Real-ism," in *Theory and Event* 18, no. 2 (2015); reprinted as "The Gravity of Melancholia: A Critique of Speculative Realism," in *Politics, Theory, and Film: Critical Encounters with Lars von Trier*, edited by Bonnie Honig and Lori Marso, 389–412 (Oxford: Oxford University Press, 2016); repro-duced by permission of Oxford University Press. The section in chapter 1 entitled "The Effanineffable Name of Language" first appeared in substan-tially different form as "The Monolingualism of the Human," in *Substance* 43, no. 2 (2014): 83–99; reproduced courtesy of the University of Wiscon-sin Press.

NOTES

INTRODUCTION

1. Kafka, "Report to an Academy," 80.

2. Simons and Chabris, "Gorillas in Our Midst," 1059–1074. A video of the test can be found at https://www.youtube.com/watch?v=vJG698 U2Mvo.

3. Singer, *Animal Liberation*, 6.

4. Lovgren, "Chimps, Humans 96 Percent the Same"; Freud, "Taboo of Virginity," 199; Freud, "Civilization and Its Discontents," 99.

5. Agamben, *The Open*, 37. Agamben argues that the premodern machine operated in an inverted fashion by humanizing animal life such that some humans came to signify the border between human and animal: the slave, the barbarian, the foreigner, etc.

6. Ibid., 26.

7. Ibid., 25, 26.

8. See Wolfe, *Animal Rites*, 124.

9. Baker, *Picturing the Beast*, 136.

10. Ibid., 211, 138.

11. Burt, "Review of *Zoontologies*," 168.

12. Benjamin, "Franz Kafka," 122.

13. Husserl, "Foundational Investigations," 117–131.

14. Bennett, *Vibrant Matter*, 2.

15. Ibid., 51, 2, 52.

16. Ibid., 52.

17. Ibid., 2.

18. Bentham, *Principles of Morals and Legislation*, 311.

19. Derrida, *Animal*, 28.

20. Ibid.

21. Ibid. Projecting the passivity of death onto animals, Heidegger famously declared that "only man dies." The animal merely "perishes" because death names an ostensibly unique human capability or capacity to know death "as such." See Heidegger, "The Thing," 178. As Derrida counters, however, humans no more have access to the "as such" of death than do

animals because finitude marks precisely the threshold that separates life from an inaccessible death. See Derrida, *Aporias*, 77.

22. On this necessary and irresolvable tension between univocity and equivocity, see Derrida, *Edmund Husserl's Origin of Geometry*, 105. I provide an expanded discussion of this tension in chapters 1 and 4.

23. Naas, *Derrida from Now On*, 200.

24. Pettman, *Human Error*, 203.

25. Ibid., 196.

26. Ibid., 7, 6, 205.

27. Ibid., 199; Agamben, *The Open*, 38.

28. Agamben, *The Open*, 92.

29. Meillassoux, *After Finitude*, 136.

30. Kafka, "Report to an Academy," 83.

31. Ibid., 80.

32. Meillassoux, *Time without Becoming*, 10.

33. Ibid., 20.

34. Coetzee, "Realism," 18.

35. See Makela, *Munich Secession*, 30.

36. Derrida, *Animal*, 6.

37. Derrida, "Poetics and Politics of Witnessing," 76, 77.

38. Ibid., 86.

39. Derrida, *Animal*, 4. David Wood's article "If a Cat Could Talk" is accompanied by a photo of Derrida holding a Siamese cat in his lap. The caption gives the name "Logos" to the animal, but Wood does not mention this name in the article. One of Derrida's translators has advised me that Derrida called the cat depicted in the photo Lucrèce. A number of websites give the name Logos to Derrida's cat based on a quotation from "Plato's Pharmacy": "Logos, a living animate creature, is thus also an organism that has been engendered." Yet in context Derrida is clearly not talking about his or any other cat. See Derrida, "Plato's Pharmacy," 84.

40. Derrida, *Animal*, 9.

41. Derrida, *Gift of Death*, 114.

42. Coetzee, "Exposing the Beast." Kari Weil argues similarly that, "unlike in women's studies or ethnic studies . . . those who constitute the objects of animal studies cannot speak for themselves, or at least they cannot speak any of the languages that the academy recognizes as necessary for such self-representation" (*Thinking Animals*, 4).

43. Morton, "Art in the Age of Asymmetry," 132.

44. Morton, "Here Comes Everything," 180.

45. Derrida, "Force of Law," 24.

46. On the distinction between undecidability and indecisiveness, see Rapaport, *Theory Mess*, 121–124.

47. Derrida, *On Cosmopolitanism and Forgiveness*, 23.
48. Marder, "For a Phytocentrism to Come," 249.
49. Calarco, *Zoographies*, 10.
50. Esposito, *Bíos*, 186.
51. Wolfe, *Before the Law*, 103.
52. Ibid.
53. Ibid.
54. Ibid., 86.
55. See Wolfe, *Before the Law*, 93. Wolfe acknowledges that the *à venir* does not equate with the Kantian idea, yet teleology slips in through the back door when he figures full inclusivity as either desirable but impossible or desirable and achievable. As I argue in chapter 4, the *à venir* is neither reducible to the Kantian idea nor diametrically opposed to it.
56. Ibid., 92.
57. I draw here from a similar claim advanced by Martin Hägglund that the perfect democracy is *normatively* undesirable. As I explain in chapter 4, however, I disagree with him that democracy (or anything else for that matter) is *descriptively* undesirable in the sense that we cannot desire it. See Hägglund, *Radical Atheism*.
58. Freud, "Fixation to Traumas," 285; Haraway, *When Species Meet*, 11.
59. Danielson, "Great Copernican Cliché," 1033.
60. See Schiebinger, *Nature's Body*, 80; Buffon, "Nomenclature des singes," 215. Rousseau, *Discours*, 226; See also Camper, "Organs of Speech," 139–159; Fellows and Milliken, "A Precursor of Darwin?," 112–124.
61. The "wound" narrative also does not consider popular curiosity with natural history in the nineteenth century, which prompted museum exhibits of orangutans as well as ubiquitous performances of "missing links" and "monkey men" on Victorian stages in the decades prior to the publication of Darwin's *Origin*. See Goodall, *Performance and Evolution*, 9. See also Fleming and Goodall, "Dangerous Darwinism," 259–271.
62. Haraway, *When Species Meet*, 12.
63. Lestel, *L'Animal Singulier*, 59–86. Carrie Rohman also writes of "Darwin's catastrophic blow to human privilege" in *Stalking the Subject*, 1. I would be remiss not to acknowledge that Derrida also registers his belief in the Copernican and Darwinian myths in *The Animal That Therefore I am*. That he does so in the context of an argument that insists on the nonpower of the human to erase its or any other trace, however, suggests that he does not conceive this series of wounds as belonging to a teleological movement that could successfully eradicate narcissism. See Derrida, *Animal*, 136.
64. Pettman, *Human Error*, 21.
65. Freud, "An Outline of Psychoanalysis," 61. For an expanded discussion of disavowal as it relates to the intersection of race, sexuality, and

animality, see the introduction to my *Bestial Traces*. On disavowal and ego splitting as a fundamental component of psychopathology, see Bass, *Difference and Disavowal*, 33.

66. Meillassoux, *After Finitude*, 7.

67. Carroll, *Through the Looking-Glass*, 161.

68. I thank J. Hillis Miller for alerting me to the etymology of cabbage as well as to the anthropomorphism it implies.

69. Husserl, *Cartesian Meditations*, 92.

70. Zahavi, *Husserl's Phenomenology*, 114.

71. Butler, *Gender Trouble*, 143.

72. Whitman, *Leaves of Grass*, 715.

73. Gratton, *Speculative Realism*, 3.

74. These academic publishers include University of Minnesota Press, University of Chicago Press, Duke University Press, Bloomsbury Publishing, Edinburgh University Press, and Open Humanities Press.

75. Bryant, *Democracy of Objects*, 19.

76. Gratton, *Speculative Realism*, 2, 1, 2.

77. Sontag, "Imagination of Disaster."

78. Gratton, *Speculative Realism*, 1.

79. Grusin, "Introduction," xix.

1. THE SCANDAL OF THE HUMAN: IMMANENT TRANSCENDENCY
AND THE QUESTION OF ANIMAL LANGUAGE

1. Colebrook, "Not Symbiosis, Not Now," 202.

2. Bennett, *Vibrant Matter*, ix.

3. Ibid., 120.

4. Ibid., ix.

5. *An Intermediate Greek-English Lexicon*, s.v. "*Skándalon*," accessed May 12, 2016, http://www.perseus.tufts.edu/hopper/.

6. Derrida, *Beast and the Sovereign*, 2:22.

7. Ibid.

8. Ibid., 28.

9. Ibid.

10. Ibid., 29.

11. Ibid.

12. Ibid.

13. Coetzee, "Exposing the Beast."

14. Ibid.

15. Heidegger, *Fundamental Concepts of Metaphysics*, 193.

16. Haraway, *When Species Meet*, 165.

17. Ibid., 14, 42.

18. Ibid., 165.

19. Husserl, *Cartesian Meditations*, 109.
20. Derrida, "Violence and Metaphysics," 125.
21. Ibid., 121.
22. Husserl, *Cartesian Meditations*, 92.
23. Ibid.
24. Haraway, *When Species Meet*, 274.
25. Husserl, *Ideas 1*, 58.
26. Derrida, "Rams," 160–161. For further reflections on the theme of solitude in Derrida and its association with Husserlian appresentation, see Miller, "Derrida Enisled," 101–132.
27. Pettman, *Human Error*, 20, 208.
28. Grusin, "Introduction," ix, xx.
29. Ibid., xx.
30. Ibid.
31. Ibid., xvii, ix.
32. Kant, *Lectures on Ethics*, 213.
33. Ibid.
34. Bennett, *Vibrant Matter*, 108.
35. Ibid., viii. While Bennett shares with object-oriented ontology a concern for the agency and vitality of things, she does not describe her work as object-oriented ontology. Bennett's vibrant materialism is in fact distinct from object-oriented ontology insofar as it does not position itself as anti-Kantian, and thus does not disavow subjectivity or relationality *tout court*.
36. Ibid., 100.
37. Ibid., 107.
38. Ibid., 122. See Levinas, "Paradox of Morality," 168–180. According to Levinas: "The ethical extends to all living beings. We do not want to make an animal suffer needlessly and so on" (172). He expresses this concern for animal suffering independent of the Kantian principle that animal cruelty fosters harm toward humans. Yet he does not affirm that animals have faces, which would mean that, consistent with his larger philosophical principles, animal slaughter is unethical. Kant acknowledged animal suffering, but he nevertheless acquiesced to the practice of vivisection because "it is employed for a good purpose" (*Lectures on Ethics*, 213). Levinas does not comment on vivisection, but he gives the impression that some suffering is justifiable: "Vegetarianism, for example, arises from the transference to animals of the idea of suffering. The animal suffers. It is because we, as human, know what suffering is that we can have this obligation" (172). If "human ethics" is the "prototype" according to which we should not cause unnecessary animal suffering, then carnivorism would seem to occupy the same position in Levinasian ethics as vivisection does in Kantian ethics: an unavoidable yet justifiable cruelty (ibid.).
39. Bennett, *Vibrant Matter*, 13.

40. Dickinson, *Complete Poems*, 116.

41. Ibid., 165.

42. Naas, *Derrida from Now On*, 188.

43. Ibid., 193.

44. Ibid., 192.

45. Derrida, *Voice and Phenomenon*.

46. Ibid., 89.

47. Ibid., 88.

48. Derrida, *H. C. for Life*, 115. DeArmitt, *Right to Narcissism*, 3.

49. Derrida, "No *One* Narcissism," 199.

50. Zahavi, "Empathy," 289.

51. Kant, *Lectures on Ethics*, 212.

52. Ibid., 51, 2, 52.

53. Ibid., 52.

54. Derrida, "Eating Well," 285.

55. Lestel, "How Chimpanzees Have Domesticated Humans," 12–15. On the topic of how these experiments produce talking apes, see Chrulew, "Philosophical Ethology of Dominique Lestel," 17–44.

56. Lestel, *L'animal singulier*, 120; my translation.

57. Lestel, "How Chimpanzees Have Domesticated Humans," 14.

58. Pollick and de Waal, "Ape Gestures and Language Evolution," 8184–8189.

59. Lestel, *L'animal singulier*, 119; my translation.

60. Derrida, *Monolingualism of the Other*, 2.

61. Ibid., 25.

62. Lacan, *Psychoses*, 167. A recent article published in *Nature* reconfirms von Frisch's theory. See Riley et al., "Flight Paths of Honeybees," 205–207.

63. de Man "Autobiography as De-Facement," 70.

64. Pepperberg, *Alex Studies*, 41.

65. Ibid., 44.

66. Pepperberg, *Alex and Me*.

67. Terrace et al., "Can an Ape Create a Sentence?," 900.

68. Smith, "Thinking Bird or Just Another Birdbrain?"

69. Derrida, *Rogues*, 7.

70. Pepperberg, *Alex Studies*, 41.

71. Ibid.

72. Ibid., 44.

73. Ibid.

74. Lestel, "Mirror Effects," 55.

75. Derrida, *Edmund Husserl's Origin of Geometry*, 105.

76. Ibid., 103.

77. Derrida, *Animal*, 37.
78. T. S. Eliot, *Old Possum's Book of Practical Cats*, 2.
79. Massumi, *What Animals Teach Us About Politics*, 8.
80. Ibid., 8, 45.
81. Ibid., 26, 8.
82. Ibid., 2, 8.
83. Ibid., 8.
84. Ibid., 50.
85. Ibid., 77.
86. Ibid., 3. Derrida, *Beast and the Sovereign*, 2:74.
87. Defoe, *Robinson Crusoe*, 152.
88. Derrida, *Beast and the Sovereign*, 2:35.
89. Lestel, *L'animal singulier*, 139; my translation.

2. SOVEREIGN SILENCE: THE DESIRE FOR ANSWERING SPEECH

1. Kafka, "Silence of the Sirens," 128.
2. I discuss these racist representations in the introduction to my *Bestial Traces*, 1–7.
3. Coetzee, *Foe*, 147.
4. Ibid., 80.
5. MacLeod, "'Do we of necessity become puppets in the story?' Or Narrating the World," 1–18.
6. Attridge, "Oppressive Silence," 226.
7. Ibid., 152.
8. Ibid., 153.
9. See Freud, *Interpretation of Dreams*, 296–299.
10. Coetzee, *Foe*, 152.
11. Ibid., 154.
12. Ibid., 152.
13. Ibid., 141.
14. Ibid., 142.
15. Ibid., 157.
16. Ibid., 8.
17. Ibid., 59.
18. Ibid., 118.
19. Quayson, *Aesthetic Nervousness*, 164.
20. Ibid., 150, 149.
21. Kalpana Rahita Seshadri argues similarly that we should avoid reducing silence either to an effect of repression or to a "pristine state" exterior to language. See her *HumAnimal*, 33.
22. Derrida, *Monolingualism of the Other*, 2, 23.
23. Ibid.

24. Ibid., 39, 40.

25. Ibid., 1.

26. Ibid., 25.

27. Husserl, *Logical Investigations*, 1:191.

28. Derrida, *Voice and Phenomenon*, 42.

29. We need only replace the word *bodies* with *words* in Coetzee's phrase to grasp this impossible idiomaticity: *This is a place where words are their own signs.* How could words be their own signs? How could any word belong fully to itself? How could it sign anything except on the condition that it exist in relation to other signs whose alterity deprives it of its ownness?

30. Ibid., 88.

31. Derrida, *Monolingualism of the Other*, 22.

32. Derrida refers explicitly to Husserl's notion of analogical appresentation in a number of texts in addition to *Violence and Metaphysics* and *Voice and Phenomenon*. See "Rams," 135–164; and *Gift of Death*.

33. Derrida, *Voice and Phenomenon*, 65.

34. Derrida does not employ the concept of "citationality" in *Voice and Phenomenon*, but his discussion of Husserl anticipates the circulation of this term in later texts such as "Signature Event Context." See Derrida, "Signature Event Context," 307–330.

35. Ibid., 69.

36. Ibid.

37. Coetzee, *Doubling the Point*, 248.

38. Ibid.

39. Ibid., 248, 249.

40. Parry, "Speech and Silence," 154, 158.

41. Derrida, *Monolingualism of the Other*, 64.

42. Ibid., 65.

43. Coetzee, *Foe*, 157.

44. This opacity of meaning is further muddied by the puzzling dream logic that conflates the individual events that brought Cruso, Friday, and Susan to the island. In one of his conflicting stories, Cruso claims that he and Friday were the sole survivors of a shipwreck. Although Susan was cast adrift in a rowboat along with her dead captain when the crew of their ship mutinied, the final section of the novel imagines Susan, Friday, and the captain entombed within the same wreckage.

45. Coetzee, *Foe*, 22.

46. Ibid., 69.

47. Ibid., 98, 119.

48. Ibid., 146.

49. Ibid., 198.

50. Defoe, *Robinson Crusoe*, 213.
51. Coetzee, *Foe*, 56.
52. Brown, "In the 'folds of our own discourse,'" 186.
53. Coetzee, *Foe*, 61.
54. Foucault, *History of Sexuality*, 1:101.
55. Brown, "In the 'folds of our own discourse,'" 189.
56. Chesnutt, *Conjure Woman*.
57. Ibid., 163, 166.
58. Ibid., 160.
59. Ibid., 170, 169.
60. Ibid., 172.
61. Derrida, *Beast and the Sovereign*, 1:32.
62. Derrida, *Rogues*, 100.
63. I say "almost always" not to imply that an individual act of slave resistance can overturn racial hierarchies altogether. Yet if a slave were to commit an extreme act of resistance, such as killing the master, she can reasonably be said to have gained the upper hand, at least fleetingly. Moreover, the impossible purity of sovereignty *as such*, should not be collapsed onto the *experience* of slavery in which the master's authority is felt *as if* it were absolute.
64. Sundquist, *To Wake the Nations*, 390.
65. Chesnutt, *Conjure Woman*, 171.
66. Ibid., 162, 160.
67. Sundquist, *To Wake the Nations*, 392.
68. Chesnutt, *Conjure Woman*, 162.
69. Washington, *Up from Slavery*, 5.
70. Ibid. Washington goes on to explain that "this news was usually gotten from the coloured man who was sent to the post-office for the mail. In our case the post-office was about three miles from the plantation, and the mail came once or twice a week. The man who was sent to the office would linger about the place long enough to get the drift of the conversation from the group of white people who naturally congregated there, after receiving their mail, to discuss the latest news. The mail-carrier on his way back to our master's house would as naturally retail the news that he had secured among the slaves, and in this way they often heard of important events before the white people at the 'big house,' as the master's house was called."
71. Bone, *Down Home*, 83.
72. Ibid., 172.
73. "The Dumb Witness" was accepted for publication in 1897, but ultimately was not published during Chesnutt's lifetime. A modified version was incorporated into Chesnutt's novel, *The Colonel's Dream* (1905). For more

on the complicated publication history of Chesnutt's story, see Richard
Broadhead's introduction to *The Conjure Woman and Other Conjure Tales*, 18.

   74. Coetzee, *Foe*, 38.

   75. Derrida, *Beast and the Sovereign*, 2:9.

   76. Ibid., 8.

   77. Derrida, *Voice and Phenomenon*, 56.

   78. Coetzee, *Foe*, 30.

   79. This "yes" is related to Derrida's notion of the *arrivant*, which
signifies whoever or whatever arrives (*ce qui arrive*). Even if we say "no" to
this *arrivant* because it threatens injury or harm, this negation nevertheless
presupposes a prior "yes," an originary vulnerability and exposure to a
world with others. See Derrida, *Aporias*, 33–34; and Derrida, "A Number
of Yes," 231–240.

3. THE GRAVITY OF MELANCHOLIA:
A CRITIQUE OF SPECULATIVE REALISM

   1. Derrida, "Circumfession," 134.

   2. *Melancholia*, directed by Lars von Trier (Zentropa, 2011), theatrical
release, and *Gravity*, directed by Alfonso Cuarón (Warner Bros., 2013), theatrical release.

   3. Freud, "Mourning and Melancholia," 245.

   4. Harman, "Well-Wrought Broken Hammer," 187.

   5. Meillassoux, *After Finitude*, 119.

   6. Bryant, *Democracy of Objects*, 246.

   7. Cuarón and Cuarón, *Gravity*.

   8. Bryant, *Democracy of Objects*, 32.

   9. Meillassoux, *After Finitude*, 128.

   10. Harman, *Quadruple Object*, 5.

   11. Ibid., 3.

   12. Bogost, *Alien Phenomenology*, 21.

   13. Bryant, *Democracy of Objects*, 22; Husserl, *Ideas 1*, 35.

   14. Husserl, *Cartesian Meditations*, 30, 120.

   15. Husserl, *Crisis of European Sciences*, 185.

   16. Harman, "Well-Wrought Broken Hammer," 185.

   17. Ibid.

   18. Harman, *Quadruple Object*, 46.

   19. Ibid., 45.

   20. Ibid.

   21. Spinoza, "Letter 58," 909.

   22. Harman, "Aesthetics as First Philosophy."

   23. Ibid.

24. Bennett, "Systems and Things," 226.

25. Bogost, *Alien Phenomenology*, 8.

26. Harman, *Quadruple Object*, 46.

27. Derrida, *Animal*, 34.

28. Bryant, *Democracy of Objects*, 19.

29. Gratton, *Speculative Realism*, 119.

30. Bogost, *Alien Phenomenology*, 11.

31. Levinas, *Totality and Infinity*.

32. Harman, "Aesthetics as First Philosophy."

33. Derrida, "Violence and Metaphysics," 125.

34. Bennett, "Systems and Things," 230.

35. Bryant, *Democracy of Objects*, 40.

36. Harman, *Quadruple Object*, 63.

37. See my "Slavery's Bestiary: Joel Chandler Harris's *Uncle Remus Tales*," in *Bestial Traces*, 50–73.

38. Derrida, *Rogues*, 11.

39. Ibid., 195.

40. Ibid., 199.

41. Bogost, *Alien Phenomenology*, 22, 21.

42. As Derrida argues, "absolute power can be figured by the grandeur of the grandest but also by smallness, arch-smallness, the absolute diminution of the smallest. . . . The political power that is today trying to make its sovereignty prevail thanks to its economic and techno-scientific resources (I was mentioning last time the satellites of worldwide surveillance, whose information is sometimes bought by the United Slates) does so through the refinement of what are now called *nanotechnologies*" (*Beast and the Sovereign*, 1:257–258).

43. Freud, "Mourning and Melancholia," 254, 255.

44. While I subscribe to Derrida's view that mourning and melancholia are not absolutely distinct—that is, the death of the other engenders an irremediable loss of the world, thereby forcing us to "carry the world of the other . . . after the end of the world"—some forms of melancholia are clearly more livable than others, more open to a future that says "yes" to mourning even if this affirmation must remain equivocal. Mourning may be interminable, but a mourning that says "yes" to loss without any pang of regret would be, in the final analysis, indistinguishable from the disavowal of loss that conditions melancholia in the first place. See Derrida, "Rams," 140.

45. Harman, *Third Table*, 15.

46. Ibid., 12.

47. Norris, *Philosophy Outside-In*, 195.

48. Bogost, *Alien Phenomenology*, 4.

49. Freud, "Mourning and Melancholia," 254.

50. Ibid., 255, 254.

51. Phillips, *On Kissing*, 76.

52. Derrida, "Some Statements and Truisms," 73.

53. Morton, "Here Comes Everything," 163.

54. Ibid., 170.

55. Bryant, *Democracy of Objects*, 29.

56. Harman, "Well-Wrought Broken Hammer," 200.

57. Ibid., 185.

58. Faulkner, *As I Lay Dying*, 160.

59. Coetzee, *Foe*, 152.

60. Meillassoux, *After Finitude*, 124.

61. Ibid., 121.

62. Ibid., 10; Kant, "Preface to the Second Edition," 112.

63. Meillassoux, *After Finitude*, 14, 13.

64. Ibid. The original French states that the correlationist codicil is "discretèment [discreetly] placé en bout de phrase." The English translation introduces the incorrect "discretely." See Meillassoux, *Après la finitude*, 30.

65. Ibid., 15.

66. Ibid.

67. Husserl, *Crisis of European Sciences*, 53, 48. See also Zahavi, *Husserl's Phenomenology*, 128. Zahavi observes that Husserl criticizes "certain elements in the inflated self-understanding of science" rather than reject science altogether.

68. Meillassoux, *After Finitude*, 17.

69. Ibid., 115, 116.

70. Ibid., 117.

71. Ibid., 118.

72. Ibid., 136.

73. Freud, "Fixation to Traumas," 285.

74. *Maimonides, Guide of the Perplexed*, 192, cited in Brague, "Geocentrism as a Humiliation for Man," 198.

75. Dante, *Divine Comedy*, 185.

76. Galileo Galilei, *Galileo on the World Systems*, 90.

77. Kant, "Preface to the Second Edition," 113.

78. Ibid., 110.

79. Arendt, *On Revolution*, 42.

80. Ibid., 45. Arendt suggests that this shift in sense from restoration to transformation explains why no less a revolutionary than Thomas Paine could characterize the French and American revolutions as "counterrevolutions" (in

a positive sense) that declared the inalienable political rights of all men by virtue of birth yet denied by centuries of tyranny.

81. Husserl, "Foundational Investigations," 130.

82. Ibid., 118.

83. Ibid., 123.

84. Meillassoux, *After Finitude*, 112.

85. *Another Earth*, directed by Mike Cahill (Fox Searchlight Pictures, 2011), DVD.

86. *Solaris*, directed by Andrei Tarkovsky (Criterion Collection, 2011), DVD.

87. The figure of the counter-earth can be traced to the origins of science fiction cinema in the iconic scene from Georges Méliès's *Le Voyage dans la Lune* (1902), in which a bullet-shaped space capsule carrying earthly inhabitants strikes the eye of the moon's anthropomorphized face. Humans project their own image as far away as the moon, whose vision must be immediately rendered half blind, as if to proscribe in advance the possibility of any cosmic reverse shot, thus preserving the unidirectional gaze. For an analysis of this scene from Méliès's film in terms of what this interplanetary shot/reverse shot implies for even the most intimate of intraterrestrial gazes between self and other, see Szendy, *Kant in the Land of Extraterrestrials*, 129–133.

88. Meillassoux, *After Finitude*, 112.

89. Kamuf, "Competent Fictions," 9.

90. Sontag, "Imagination of Disaster."

91. Carlsen, "The Only Redeeming Factor is the World Ending."

92. I previously discussed this scene and several others from von Trier's film in an analysis that interrogated the false choice between literal and allegorical interpretation. See my "Magic Cave of Allegory," 400–422.

93. Derrida, "Rams," 161.

4. LISTING TOWARD COSMOCRACY:
THE LIMITS OF HOSPITALITY

1. Szendy, *Kant in the Land of Extraterrestrials*, 145.

2. Whitman, *Leaves of Grass*, 715.

3. Bennett, "The Solar Judgment of Walt Whitman," 138.

4. Ibid., 131.

5. Killingsworth, *Walt Whitman and the Earth*, 23.

6. Coviello, *Intimacy in America*, 135.

7. Warner, "Whitman Drunk," 40.

8. Whitman, *Leaves of Grass*, 127.

9. Ibid.

10. Lawrence, "Whitman," 15, 16, 23.

11. Sommer, "Supplying Demand," 52.

12. Whitman, *Leaves of Grass*, 48.

13. Ibid., 52. 1855 edition.

14. Ibid., 715; ellipses in the original.

15. Ibid., 233.

16. Plato, *Republic*, 215.

17. Levinas, *Time and the Other*, 64.

18. Ibid.

19. Ibid., 65.

20. Derrida, "Violence and Metaphysics," 92.

21. Bennett, "Systems and Things," 227.

22. Ibid., 77.

23. Ibid., 32.

24. Bennett, "The Solar Judgment of Walt Whitman," 136.

25. Bogost, "Tacos, Enchiladas, Burritos, Chiles, &c."; Bogost, "Latour Litanizer."

26. In "The Poet," Emerson wrote that "bare lists of words are found suggestive, to an imaginative and excited mind" (334). The objects cataloged by OOO are deliberately chosen for their apparent incongruity, which is to say that they are chosen precisely to give the impression of not having been chosen, of having been selected at random, as if attesting to a form of nonjudgmental, "solar" judgment.

27. Bennett, "Of Material Sympathies," 249.

28. Whitman, *Leaves of Grass*, 431.

29. Erkkila, *Whitman the Political Poet*, 81–91.

30. Whitman, *Leaves of Grass*, 565.

31. Ibid., 425.

32. Ibid., 36.

33. Previously signifying a human or animal bed as well as nonhuman offspring, litter assumed an expanded sense in the nineteenth century due to its association with straw and animal waste, leading to its modern equation with disorder and debris.

34. Bennett, "The Solar Judgment of Walt Whitman," 134.

35. Whitman, *Leaves of Grass*, 716, 165.

36. Bennett, "The Solar Judgment of Walt Whitman," 134.

37. Whitman, *Leaves of Grass*, 348.

38. For more on Whitman's complex relation to feminism, see Pollak, "In Loftiest Spheres," 172–193.

39. Bennett, "The Solar Judgment of Walt Whitman," 140.

40. Cole, "Call of Things," 106, 112.

41. Latour, *Reassembling the Social*, 79.

42. Bennett, "The Solar Judgment of Walt Whitman," 137.

43. Ibid., 142.

44. Ibid., 133.

45. Ibid., 138.

46. Ibid., 142.

47. Bergson, *Matter and Memory*, 33.

48. Ibid., 66.

49. Bennett, "The Solar Judgment of Walt Whitman," 132.

50. Bergson, *Matter and Memory*, 34.

51. Ibid., 75.

52. Bennett, "The Solar Judgment of Walt Whitman," 138.

53. Peter Gratton argues persuasively that speculative realists, in particular Meillassoux and Harman, exclude temporality and thereby reinscribe the metaphysics of presence. If virtually everything is an object for Harman, then how, Gratton asks, can we account for an object such as music since it depends precisely on time? See Gratton, "Post-Deconstructive Realism," 84–90.

54. Ibid., 133.

55. Erkkila, *Whitman the Political Poet*, 102.

56. Whitman, *Notes and Fragments*, 19.

57. Cited in Folsom, "Lucifer and Ethiopia," 76.

58. Ibid., 78, 83, 81.

59. Whitman, *Notebooks and Unpublished Prose Manuscripts*, 2:160.

60. Gewirtz, "I Am With [Some of] You," 11.

61. Traubel, *With Walt Whitman in Camden*, 2:535.

62. Derrida, "Structure Sign and Play," 279.

63. Whitman, *Leaves of Grass*, 88.

64. Derrida, *Rogues*, 48.

65. Wallace, "E Unibus Pluram," 151–194.

66. Derrida, *Rogues*, 36.

67. Ibid.

68. Ibid., 54.

69. Ibid., 29.

70. Whitman, *Leaves of Grass*, 457.

71. Ibid., 458.

72. Ibid., 460.

73. Ibid., 458.

74. Ibid., 458.

75. Ibid., 457.

76. Ibid., 461. This identification of America with novelty and futurity in opposition to European morbidity is one of the quintessential gestures of

American exceptionalism. The myth of "virgin land," for instance, not only licensed the extermination of native peoples, but equated Europe with temporality and America with timeless space, thus exempting the latter from "the human experience of birth, death, and rebirth" (Noble, *Death of a Nation*, xl). In this regard, the "living present" of Whitman's new world, a present that contains both the past and the future, echoes George Berkeley's "Verses on the Prospect of Planting Arts and Learning in America" (1728), which similarly characterizes America as "time's noblest offspring," a progeny that escapes European "decay" (234).

77. Ibid.
78. Ibid., 459.
79. Ibid., 460.
80. Ibid., 457.
81. Marder, *Plant-Thinking*, 51.
82. Morton, "Here Comes Everything," 163.
83. Marder, "For a Phytocentrism to Come," 249.
84. For an expanded critique of Marder, see my "Races."
85. Derrida, "Politics and Friendship," 181.
86. Naas, *Derrida from Now On*, 22.
87. Bennington, "In Dignity," 7.
88. Derrida, *Rogues*, 83.
89. Derrida, *Voyous*, 122, cited in Bennington, "In Dignity," 12.
90. Bennington, "In Dignity," 12.
91. Ibid., 13.
92. Derrida, *Edmund Husserl's Origin of Geometry*, 105.
93. Ibid., 102.

94. "Cette réduction doit être indéfiniment recommencée, car le langage ne peut ni ne doit se maintenir sous la protection de l'univocité" (Derrida, "Introduction," in Husserl, *L'origine de la géometrie*, 104).

95. Lawlor, *Derrida and Husserl*, 122.
96. Derrida, *Rogues*, 74.
97. Derrida, *Edmund Husserl's Origin of Geometry*, 28.
98. Hägglund, *Radical Atheism*, 169.
99. Ibid., 32.
100. Ibid., 202.
101. Ibid., 170.
102. Ibid.

103. In a critique of Hägglund, Naas asks, "Isn't it possible to desire and yet not desire something? Might it not make more sense to talk about an aporia or denegation of desire rather than a so-called or purported desire that dissimulates a real one?" ("An Atheism," 61).

104. Hägglund, *Radical Atheism*, 204.

105. Ibid., 169.

106. Ibid., 203.

107. Ibid.

108. See Derrida, "Signature Event Context," 324.

109. Lacan, "Direction of the Treatment," 250–310; Kojève, *Introduction to the Reading of Hegel.*

110. Kojève, *Introduction to the Reading of Hegel*, 6; Lacan, "Direction of the Treatment," 282.

111. Derrida, *Animal*, 28.

112. For an analysis of this undesirability in relation to Coetzee's *Disgrace*, see my "Ashamed of Shame" in *Bestial Traces*, 113–146.

113. Attridge, "Age of Bronze," 98–121.

114. Whitman, *Leaves of Grass*, 368.

115. Ibid.

116. Killingsworth, *Walt Whitman and the Earth*, 11.

117. Whitman, *Leaves of Grass*, 368.

118. Ibid., 369.

119. Ibid.; Killingsworth, *Walt Whitman and the Earth*, 19.

120. Whitman, *Leaves of Grass*, 369

121. Ibid.

122. Derrida, *Rogues*, 10.

123. Ibid., 37.

124. Calarco, "Toward an Agnostic Animal Ethics," 78.

125. Ibid., 81.

126. Ibid.

127. Ibid., 79, 80.

128. Ibid., 77.

129. Ibid., 80.

130. Fleming and O'Carroll, "Paganism."

131. Foucault, *Order of Things*, 422.

132. An Intermediate Greek-English Lexicon, s.v. "ἀπορία" (aporia), accessed May 15, 2016, http://www.perseus.tufts.edu/hopper.

133. Whitman, *Leaves of Grass*, 458.

Agamben, Giorgio. *The Open: Man and Animal*. Translated by Kevin Attell. Stanford, CA: Stanford University Press, 2004.

Arendt, Hannah. *On Revolution*. New York: Penguin, 1990.

Attridge, Derek. "Age of Bronze, State of Grace: Music and Dogs in Coetzee's *Disgrace*." *Novel: A Forum on Fiction* 34, no. 1 (2000): 98–121.

———. "Oppressive Silence: J. M. Coetzee's *Foe* and the Politics of the Canon." In *Decolonizing Tradition: New Views of Twentieth-century "British" Literary Canons*, edited by Karen Lawrence, 212–238. Champaign: University of Illinois Press, 1992.

Baker, Steve. *Picturing the Beast: Animals, Identity, and Representation*. Champaign: University of Illinois Press, 2001.

Bass, Alan. *Difference and Disavowal: The Trauma of Eros*. Stanford, CA: Stanford University Press, 2000.

Benjamin, Walter. "Franz Kafka: On the Anniversary of His Death." In *Illuminations*, translated by Harry Zohn, 111–140. New York: Schocken Books, 2007.

Bennett, Jane. "Of Material Sympathies, Paracelsus, and Whitman." In *Material Ecocriticism*, edited by Serenella Iovino and Serpil Oppermann, 239–252. Bloomington: Indiana University Press, 2014.

———. "The Solar Judgment of Walt Whitman." In *A Political Companion to Walt Whitman*, edited by John Seery, 131–146. Lexington: University of Kentucky Press, 2011.

———. "Systems and Things: A Response to Graham Harman and Timothy Morton." *New Literary History* 43, no. 2 (2012): 225–233.

———. *Vibrant Matter: A Political Ecology of Things*. Durham, NC: Duke University Press, 2010.

Bennington, Geoffrey. "In Dignity: Worthy of the Name of Michael Naas." Unpublished paper. Accessed May 12, 2016. https://www.academia.edu/221872/In_Dignity.

Bentham, Jeremy. *An Introduction to the Principles of Morals and Legislation*. Oxford: Clarendon Press, 1876.

Bergson, Henri. *Matter and Memory*. Translated by Nancy Margaret Paul and
    W. Scott Palmer. New York: Zone Books, 1991.
Berkeley, George. "Verses On the Prospect of Planting Arts and Learning
    in America." In *The Works of George Berkeley*, vol. 3, 233–234. London:
    J. F. Dove, 1820.
Bogost, Ian. *Alien Phenomenology, or What It's Like to Be a Thing*. Minneapolis:
    University of Minnesota Press, 2012.
———. "Latour Litanizer: Generate Your Own Latour Litanies." Accessed
    May 6, 2016. http://bogost.com/writing/blog/latour_litanizer/.
———. "Tacos, Enchiladas, Burritos, Chiles, &c.: A Mexican Food Index to
    Alien Phenomenology." Bogost.com (blog). Accessed May 6, 2016. http://
    bogost.com/writing/blog/tacos_enchiladas_burritos_chil/.
Bone, Robert. *Down Home: Origins of the Afro-American Short Story*. New
    York: Columbia University Press, 1975.
Brague, Rémi. "Geocentrism as a Humiliation for Man." *Medieval Encounters*
    3, no. 3 (1997): 187–210.
Brown, Wendy. "In the 'folds of our own discourse': The Pleasures and Free-
    doms of Silence." *University of Chicago Law School Roundtable* 3, no. 1
    (1996): 185–197.
Bryant, Levi R. *The Democracy of Objects*. Ann Arbor, MI: Open Humanities
    Press, 2011.
Buffon, Georges-Louis Leclerc, Comte de. *Oeuvres complètes de Buffon*. Paris:
    Baudouin Frères et N. Delangle, 1826.
Burt, Jonathan. "Review of *Zoontologies: The Question of the Animal*, edited by
    Cary Wolfe, and Cary Wolfe, *Animal Rites: American Culture, the Discourse
    of Species, and Posthumanist Theory*." *Society and Animals* 13, no. 2 (2005):
    167–170.
Butler, Judith. *Gender Trouble: Feminism and the Subversion of Identity*. New
    York: Routledge, 1990.
Calarco, Matthew. "Toward an Agnostic Animal Ethics." In *The Death of the
    Animal*, edited by Paola Cavalieri, 73–84. New York: Columbia University
    Press, 2009.
———. *Zoographies: The Question of the Animal from Heidegger to Derrida*. New
    York: Columbia University Press, 2008.
Camper, Peter. "Account of the Organs of Speech of the Orang Outang."
    Philosophical Transactions of the Royal Society of London, no. 69 (1779):
    139–159.
Carlsen, Per Juul. "The Only Redeeming Factor Is the World Ending."
    FILM, no. 72 (2011): 5–8. Accessed May 6, 2016. http://www.dfi.dk/
    Service/English/News-and-publications/FILM-Magazine/Artikler
    -fra-tidsskriftet-FILM/72/The-Only-Redeeming-Factor-is-the-World
    -Ending.aspx.

Carroll, Lewis. *Through the Looking-Glass*. In *Alice's Adventures in Wonderland and Through the Looking-Glass*, 111–241. New York: Penguin, 1998.

Chesnutt, Charles. *The Conjure Woman and Other Conjure Tales*. Durham, NC: Duke University Press, 1996.

Chrulew, Matthew. "The Philosophical Ethology of Dominique Lestel." *Angelaki: Journal of the Theoretical Humanities* 19, no. 3 (2014): 17–44.

Coetzee, J. M. *Doubling the Point: Essays and Interviews*, edited by David Attwell. Cambridge, MA: Harvard University Press, 1992.

———. "Exposing the Beast: Factory Farming Must Be Called to the Slaughterhouse." *Sydney Morning Herald*, February 22, 2007. Accessed May 6, 2016. http://www.smh.com.au/articles/2007/02/21/1171733846249.html?page=fullpage.

———. *Foe*. London: Penguin, 2010.

———. "Realism." In *Elizabeth Costello*, 1–34. New York: Viking, 2003.

Cole, Andrew. "The Call of Things: A Critique of Object-Oriented Ontologies." *Minnesota Review*, no. 80 (2013): 106–118.

Colebrook, Claire. "Not Symbiosis, Not Now: Why Anthropogenic Change Is Not Really Human," *Oxford Literary Review*, no. 34 (2012): 185–209.

Coviello, Peter. *Intimacy in America: Dreams of Affiliation in Antebellum Literature*. Minneapolis: University of Minnesota Press, 2005.

Cuarón, Alfonso, and Jonás Cuarón. *Gravity*, unspecified draft, May 29, 2012. Accessed May 6, 2016. http://www.screenplaydb.com/film/scripts/gravity/.

Danielson, Dennis R. "The Great Copernican Cliché." *American Journal of Physics* 69, no. 10 (2001): 1029–1035.

Dante. *The Divine Comedy*. Translated by Henry Wadsworth Longfellow. London: Routledge, 1867.

de Man, Paul. "Autobiography as De-Facement." In *The Rhetoric of Romanticism*, 67–81. New York: Columbia University Press, 1984.

DeArmitt, Pleshette. *The Right to Narcissism: A Case for an Im-Possible Self-Love*. New York: Fordham University Press, 2014.

Defoe, Daniel. *Robinson Crusoe*. New York: Penguin, 1985.

Derrida, Jacques. *The Animal That Therefore I Am*. Translated by David Wills. New York: Fordham University Press, 2008.

———. *Aporias*. Translated by Thomas Dutoit. Stanford, CA: Stanford University Press, 1993.

———. *The Beast and the Sovereign*. Vol. 2. Edited by Michel Lisse, Marie-Louise Mallet, and Ginette Michaud, translated by Geoffrey Bennington. Chicago: University of Chicago Press, 2011.

———. "Circumfession." In *Jacques Derrida*, by Geoffrey Bennington and Jacques Derrida. Translated by Geoffrey Bennington, 3–315 (Chicago: University of Chicago Press, 1999).

———. "'Eating Well,' or The Calculation of the Subject." Translated by Peter Connor and Avital Ronell. In *Points . . . Interviews, 1974–1994*, edited by Elizabeth Weber, 255–287. Stanford, CA: Stanford University Press, 1995.

———. *Edmund Husserl's Origin of Geometry: An Introduction*. Translated by John P. Leavey. Lincoln: University of Nebraska Press, 1989.

———. "The Force of Law: 'The Mystical Foundation of Authority.'" Translated by Mary Quaintance. In *Deconstruction and the Possibility of Justice*, edited by Drucilla Cornell and Michel Rosenfeld, 3–63. New York: Routledge, 1992.

———. *The Gift of Death*. Translated by David Wills. Chicago: University of Chicago Press, 1995.

———. *H. C. For Life, That Is to Say. . . .* Translated by Laurent Milesi and Stefan Herbrechter. Stanford, CA: Stanford University Press, 2006.

———. *The Monolingualism of the Other, or The Prosthesis of Origin*. Translated by Patrick Mensah. Stanford, CA: Stanford University Press, 1998.

———. "A Number of Yes." Translated by Brian Holmes. In *Psyche: Inventions of the Other*, vol. 2, edited by Peggy Kamuf and Elizabeth Rottenberg, 231–240. Stanford, CA: Stanford University Press.

———. *On Cosmopolitanism and Forgiveness*. Translated by Mark Dooley and Michael Hughes. New York: Routledge, 2001.

———. "Plato's Pharmacy." In *Dissemination*, translated by Barbara Johnson, 67–186. New York: Continuum, 2004.

———. "Poetics and Politics of Witnessing." In *Sovereignties in Question: The Poetics of Paul Celan*, edited by Thomas Dutoit and Outi Pasanen, 65–96. New York: Fordham University Press, 2005.

———. "Politics and Friendship." In *Negotiations: Interventions and Interviews, 1971–2001*, translated by Robert Harvey, edited by Elizabeth Rottenberg, 147–198. Stanford, CA: Stanford University Press, 2002.

———. "Rams: Uninterrupted Dialogue—Between two Infinities, the Poem." Translated by Thomas Dutoit and Philippe Romanski. In *Sovereignties in Question: The Poetics of Paul Celan*, edited by Thomas Dutoit and Outi Pasanen, 135–164. New York: Fordham University Press, 2005.

———. *Rogues: Two Essays on Reason*. Translated by Pascale-Anne Brault and Michael Naas. Stanford, CA: Stanford University Press, 2005.

———. "Signature Event Context." In *Margins of Philosophy*, translated by Alan Bass, 307–330. Chicago: University of Chicago Press, 1982.

———. "Some Statements and Truisms About Neologisms, Newisms, Postisms, Parasitisms, and Other Small Seismisms." Translated by Anne Tomiche. In *The States of "Theory": History, Art, and Critical Discourse*, edited by David Carroll, 63–94. New York: Columbia University Press, 1990.

————. "Structure Sign and Play in the Discourse of the Human Sciences." In *Writing and Difference*, translated by Alan Bass, 278–293. Chicago: University of Chicago Press, 1987.

————. "There is No *One* Narcissism." Translated by Peggy Kamuf. In *Points . . . Interviews, 1974–1994*, edited by Elizabeth Weber, 196–215. Stanford, CA: Stanford University Press, 1995.

————. *Voice and Phenomenon: Introduction to the Problem of the Sign in Husserl's Phenomenology*. Translated by Leonard Lawlor. Evanston, IL: Northwestern University Press, 2011.

————. "Violence and Metaphysics: An Essay on the Thought of Emmanuel Levinas." In *Writing and Difference*, translated by Alan Bass, 79–153. Chicago: University of Chicago Press, 1978.

Dickinson, Emily. *The Complete Poems of Emily Dickinson*. Edited by Thomas H. Johnson. Boston, MA: Little, Brown and Company, 1960.

Eliot, T. S. *Old Possum's Book of Practical Cats*. New York: Harcourt Brace, 1982.

Emerson, Ralph Waldo, "The Poet." In *Selected Writings of Ralph Waldo Emerson*, 325–349. New York: Signet, 2003.

Erkkila, Betsy. *Whitman the Political Poet*. Oxford: Oxford University Press, 1989.

Esposito, Roberto. *Bíos: Biopolitics and Philosophy*. Translated by Timothy Campbell. Minneapolis: University of Minnesota Press, 2008.

Faulkner, William. *As I Lay Dying*. New York: Vintage, 2004.

Fellows, Otis, and Stephen Milliken. "A Precursor of Darwin?" In *Buffon*, 112–124. New York: Twayne Publishers, 1972.

Fleming, Chris, and Jane Goodall. "Dangerous Darwinism." *Public Understanding of Science* 11, no. 3 (2002): 259–271.

Fleming, Chris, and John O'Carroll. "Paganism: Promising Promises and Resentful Results." In *Anthropoetics* 21, no. 2 (Spring 2016). Accessed May 12, 2016. http://www.anthropoetics.ucla.edu/ap2102/2102Fleming.htm.

Folsom, Ed. "Lucifer and Ethiopia: Whitman, Race, and Poetics Before the Civil War and After." In *A Historical Guide to Walt Whitman*, edited by David S. Reynolds, 45–96. Oxford: Oxford University Press, 2000.

Foucault, Michel. *The History of Sexuality*. Vol. 1. Translated by Robert Hurley. New York: Pantheon, 1978.

————. *The Order of Things*. New York: Routledge, 2002.

Freud, Sigmund. "Civilization and Its Discontents." In *Civilization, Society, and Religion*, vol. 12 of *Sigmund Freud*, translated by James Strachey, 64–148. New York: Vintage, 2001.

————. "Fixation to Traumas—The Unconscious." In *Introductory Lectures on Psychoanalysis*, vol. 16 of *The Standard Edition*, translated by James Strachey, 273–285. London: Hogarth Press, 1963.

———. *The Interpretation of Dreams*. Translated by James Strachey. New York: Basic Books, 2010.

———. "Mourning and Melancholia." In *On the History of the Post Psychoanalytic Movement*, vol. 14 of *The Standard Edition*, translated by James Strachey, 243–258. London: Hogarth Press, 1975.

———. "An Outline of Psychoanalysis." In *Five Lectures on Psycho-Analysis, Leonardo da Vinci and Other Works*, vol. 11 of *The Standard Edition*, translated by James Strachey, 1–64. London: Hogarth, 1973.

———. "The Taboo of Virginity." In *Pre-psycho-analytic Publications and Unpublished Drafts*, vol. 1 of *The Standard Edition*, translated by James Strachey, 191–208. London: Hogarth, 1973.

Galilei, Galileo. *Galileo on the World Systems: A New Abridged Translation and Guide*. Translated by Maurice Finocchiaro. Berkeley: University of California Press, 1997.

Gewirtz, Isaac. "'I Am With [Some of] You.'" In *"I Am With You": Walt Whitman's Leaves of Grass*, 10–47. New York: New York Public Library, 2005.

Goodall, Jane. *Performance and Evolution in the Age of Darwin: Out of the Natural Order*. New York: Routledge, 2002.

Gratton, Peter. "Post-Deconstructive Realism: It's about Time." *Speculations: A Journal of Speculative Realism*, no. 4 (2013): 84–90.

———. *Speculative Realism: Problems and Prospects*. London: Bloomsbury, 2014.

Grusin, Richard. "Introduction." In *The Nonhuman Turn*, edited by Richard Grusin, vii–xxix. Minneapolis: University of Minnesota Press, 2015.

Hägglund, Martin. *Radical Atheism: Derrida and the Time of Life*. Stanford, CA: Stanford University Press, 2008.

Haraway, Donna. *When Species Meet*. Minneapolis: University of Minnesota Press, 2008.

Harman, Graham. "Aesthetics as First Philosophy: Levinas and the Non-Human." Naked Punch (website). Accessed May 6, 2016. http://www.nakedpunch.com/articles/147.

———. *The Quadruple Object*. Winchester, England: Zero Books, 2011.

———. *The Third Table*. Ostfildern, Germany: Hatje Cantz Verlag, 2012.

———. "The Well-Wrought Broken Hammer: Object-Oriented Literary Criticism." *New Literary History* 43, no. 2 (2012): 183–203.

Heidegger, Martin. *The Fundamental Concepts of Metaphysics*. Translated by William McNeill and Nicholas Walker. Bloomington and Indianapolis: Indiana University Press, 1995.

———. "The Thing." In *Poetry, Language, Thought*, translated by Albert Hofstadter, 163–186. New York: Harper and Row, 1971.

Husserl, Edmund. *Cartesian Meditations: An Introduction to Phenomenology*. Translated by Dorian Cairns. Dordrecht, The Netherlands: Kluwer Academic Publishers, 1999.

―――. *The Crisis of European Sciences and Transcendental Phenomenology: An Introduction to Phenomenological Philosophy*. Translated by David Carr. Evanston, IL: Northwestern University Press, 1970.

―――. "Foundational Investigations of the Phenomenological Origin of the Spatiality of Nature: The Originary Ark, the Earth, Does Not Move." Translated by Fred Kersten and revised by Leonard Lawlor. In *Husserl at the Limits of Phenomenology*, translated by Leonard Lawlor and Bettina Bergo, 117–131. Evanston, IL: Northwestern University Press, 2001.

―――. *Ideas 1*. Translated by F. Kersten. The Hague: Martinus Nijhoff Publishers, 1983.

―――. *L'origine de la géometrie*. Traduction et introduction par Jacques Derrida. Paris: Presses Universitaires de France, 2002.

―――. *Logical Investigations*. Vol. 1. Translated by J. N. Findlay. New York: Routledge, 2001.

Kafka, Franz. "A Report to an Academy." In *Kafka's Selected Stories*, 76–94. Translated by Stanley Corngold. New York: Norton 2007.

―――. "The Silence of the Sirens." In *Kafka's Selected Stories*, 127–128. Translated by Stanley Corngold. New York: Norton 2007.

Kamuf, Peggy. "Competent Fictions: On Belief in the Humanities." Working paper. Presented at Western Sydney University, March 2010.

Kant, Immanuel. *Lectures on Ethics*. Translated by Peter Heath. Cambridge: Cambridge University Press, 1997.

―――. "Preface to the Second Edition." In *Critique of Pure Reason*, translated by Paul Guyer, 106–124. Cambridge: Cambridge University Press, 2008.

Killingsworth, Jimmie. *Walt Whitman and the Earth: A Study in Ecopoetics*. Iowa City: University of Iowa Press, 2004.

Kojève, Alexandre. *Introduction to the Reading of Hegel*. Translated by James H. Nichols Jr. Ithaca, NY: Cornell University Press, 1980.

Lacan, Jacques. "The Direction of the Treatment and the Principles of Its Power." In *Écrits*, translated by Alan Sheridan, 250–310. New York: Routledge, 2009.

―――. *The Psychoses: The Seminar of Jacques Lacan*. Translated by Russell Grigg. Edited by Jacques-Alain Miller. New York: Routledge, 1993.

Latour, Bruno. *Reassembling the Social: An Introduction to Actor-Network-Theory*. Oxford: Oxford University Press, 2005.

Lawlor, Leonard. *Derrida and Husserl: The Basic Problem of Phenomenology*. Bloomington: Indiana University Press, 2002.

Lawrence, D. H. "Whitman." In *Bloom's Modern Critical Views: Walt Whitman*, edited by Harold Bloom, 13–26. New York: Chelsea House, 2006.

Lestel, Dominique. *L'animal singulier*. Paris: Éditions du Seuil, 2004.

———. "How Chimpanzees Have Domesticated Humans: Towards an Anthropology of Human-Animal Communication." *Anthropology Today* 14, no. 3 (1998): 12–15.

———. "Mirror Effects." *Angelaki: Journal of the Theoretical Humanities* 19, no. 3 (2014): 47–57.

Levinas, Emmanuel. "The Paradox of Morality: An Interview with Emmanuel Levinas." Translated by Andrew Benjamin and Tamra Wright. In *The Provocation of Levinas: Rethinking the Other*, edited by Robert Bernasconi and David Wood, 168–180. New York: Routledge, 1988.

———. *Time and the Other*. Translated by Richard Cohen. Pittsburgh: Duquesne University Press, 1987.

———. *Totality and Infinity: An Essay on Exteriority*. Translated by Alphonso Lingis. London: Martinus Nijhoff, 1979.

Lovgren, Stefan. "Chimps, Humans 96 Percent the Same, Gene Study Finds." *National Geographic*, August 31, 2005. Accessed May 6, 2016. http://news.nationalgeographic.com/news/2005/08/0831_050831_chimp_genes.html.

MacLeod, Lewis. "'Do we of necessity become puppets in the story?' Or Narrating the World: On Speech, Silence, and Discourse in Coetzee's *Foe*." *Modern Fiction Studies* 52, no. 1 (2006): 1–18.

Maimonides. *Guide of the Perplexed*. Translated by S. Pines. Chicago: University of Chicago Press, 1963. Cited in Rémi Brague, "Geocentrism as a Humiliation for Man." *Medieval Encounters* 3, no.3 (1997): 187–210.

Makela, Maria. *The Munich Secession: Art and Artists in Turn-of-the-Century Munich*. Princeton, NJ: Princeton University Press, 1990.

Marder, Michael. "For a Phytocentrism to Come." *Environmental Philosophy* 11, no. 2 (2014): 237–252.

———. *Plant-Thinking: A Philosophy of Vegetal Life*. New York: Columbia University Press, 2013.

Massumi, Brian. *What Animals Teach Us About Politics*. Durham, NC: Duke University Press, 2014.

Meillassoux, Quentin. *After Finitude: An Essay on the Necessity of Contingency*. Translated by Ray Brassier. New York: Continuum, 2008.

———. *Après la finitude: essai sur la nécessité de la contingence*. Paris: Édition du Seuil, 2006.

———. *Time without Becoming*. Sesto San Giovanni, Italy: Mimesis International, 2014.

Miller, J. Hillis. "Derrida Enisled." In *For Derrida*, 101–132. New York: Fordham University Press, 2009.

Morton, Timothy. "Art in the Age of Asymmetry: Hegel, Objects, Aesthetics." *Evental Aesthetics* 1, no. 1 (2012): 121–142.

———. "Here Comes Everything: The Promise of Object-Oriented Ontology." *Qui Parle* 19, no. 2 (2011): 163–190.

Naas, Michael. "An Atheism That *Dieu merci!* Still Leaves Something to be Desired." *New Centennial Review* 9, no. 1 (2009): 45–68.

———. *Derrida from Now On.* New York: Fordham University Press, 2008.

Noble, David. *Death of a Nation: American Culture and the End of Exceptionalism.* Minneapolis: University of Minnesota Press, 2002.

Norris, Christopher. *Philosophy Outside-In: A Critique of Academic Reason.* Edinburgh: Edinburgh University Press, 2013.

Parry, Benita. "Speech and Silence in the Fictions of J. M. Coetzee." In *Writing South Africa: Literature, Apartheid, and Democracy, 1970–1995*, edited by Derek Attridge and Rosemary Jolly, 149–165. Cambridge: Cambridge University Press, 1998.

Pepperberg, Irene. *Alex and Me.* New York: Harper Perennial, 2009.

———. *The Alex Studies: Cognitive and Communicative Abilities of Grey Parrots.* Cambridge, MA: Harvard University Press, 2002.

Peterson, Christopher. *Bestial Traces: Race, Sexuality, Animality.* New York: Fordham University Press, 2013.

———. "The Magic Cave of Allegory: Lars von Trier's *Melancholia. Discourse.*" *Journal for Theoretical Studies in Media and Culture* 35, no. 3 (2014): 400–422.

———. "Races." In *The Edinburgh Companion to Animal Studies*, edited by Lynn Turner, Ron Broglio, and Undine Sellbach. Edinburgh: Edinburgh University Press, 2018.

Pettman, Dominic. *Human Error: Species-Being and Media Machines.* Minneapolis: University of Minnesota Press, 2011.

Phillips, Adam. *On Kissing, Tickling, and Being Bored: Psychoanalytic Essays on the Unexamined Life.* Cambridge, MA: Harvard University Press, 1993.

Plato, *The Republic.* Book 6. Translated by Tom Griffith. Cambridge: Cambridge University Press, 2003.

Pollak, Vivian R. "'In Loftiest Spheres': Whitman's Visionary Feminism." In *The Erotic Whitman*, 172–193. Berkeley: University of California Press, 2000.

Pollick, Amy S., and Frans B. M. de Waal. "Ape Gestures and Language Evolution." *Proceedings of the National Academy of Sciences of the United States of America* 104, no. 19 (2007): 8184–8189.

Quayson, Ato. *Aesthetic Nervousness: Disability and the Crisis of Representation.* New York: Columbia University Press, 2007.

Rapaport, Herman. *The Theory Mess: Deconstruction in Eclipse.* New York: Columbia University Press, 2001.

Riley, J. R. et al. "The Flight Paths of Honeybees Recruited by the Waggle Dance." *Nature*, no. 435 (2005): 205–207.

Rohman, Carrie. *Stalking the Subject: Modernism and the Animal*. New York: Columbia University Press, 2009.

Rousseau, Jean-Jacques. *Discours sur l'origine and les fondements de l'inégalité parmi les hommes*. Amsterdam: Marc Michel Rey, 1755.

Schiebinger, Londa. *Nature's Body: Gender in the Making of Modern Science*. Boston, MA: Beacon Press, 1993.

Seshadri, Kalpana Rahita. *HumAnimal: Race, Law, Language*. Minneapolis: University of Minnesota, 2012.

Simons, Daniel, and Christopher Chabris. "Gorillas in Our Midst: Sustained Inattentional Blindness For Dynamic Events." *Perception* 28 (1999): 1059–1074.

Singer, Peter. *Animal Liberation*. New York: Harper Collins, 2002.

Smith, Dinitia. "A Thinking Bird or Just Another Birdbrain?" *New York Times*, October 9, 1999. Accessed May 6, 2016. http://www.nytimes.com/1999/10/09/arts/a-thinking-bird-or-just-another-birdbrain.html?page wanted=all.

Sommer, Doris. "Supplying demand: Walt Whitman as the Liberal Self." *New Political Science* 7, no. 1 (1986): 39–61.

Sontag, Susan. "The Imagination of Disaster." *Commentary*, October 1, 1965. Accessed May 6, 2016. https://www.commentarymagazine.com/articles/the-imagination-of-disaster/.

Spinoza, Baruch. "Letter 58." In *Spinoza: The Complete Works*, translated by Samuel Shirley, 908–910. Indianapolis: Hackett Publishing, 2002.

Sundquist, Eric. *To Wake the Nations: Race in the Making of American Literature*. Cambridge, MA: Harvard University Press, 1993.

Szendy, Peter. *Kant in the Land of Extraterrestrials: Cosmopolitical Philosofictions*. New York: Fordham University Press, 2013.

Terrace, Herbert et al. "Can an Ape Create a Sentence?" *Science* 206, no. 4421 (1979): 891–902.

Traubel, Horace. *With Walt Whitman in Camden*. Vol. 2. New York: Mitchell Kennerley, 1915. Accessed May 6, 2016. http://www.whitmanarchive.org/criticism/disciples/traubel/WWWiC/2/whole.html.

Wallace, David Foster. "E Unibus Pluram: Television and U.S. Fiction." *Review of Contemporary Fiction* 13, no. 2 (1993): 151–194.

Warner, Michael. "Whitman Drunk." In *Breaking Bounds: Whitman and American Cultural Studies*, edited by Betsy Erkkila and Jay Grossman, 30–43. Oxford: Oxford University Press, 1996.

Washington, Booker T. *Up from Slavery*. New York: Barnes and Noble Publishing, 2003.

Weil, Kari. *Thinking Animals: Why Animal Studies Now?* New York: Columbia University Press, 2012.

Whitman, Walt. *Leaves of Grass.* New York: Norton, 1973.

———. *Notebooks and Unpublished Prose Manuscripts.* Edited by Edward Grier. New York: New York University Press, 1984.

———. *Notes and Fragments.* Edited by Richard Maurice Bucke. London and Ontario: A. Talbot and Co., 1899.

Wolfe, Cary. *Animal Rites: American Culture, The Discourse of Species, and Post-humanist Theory.* Chicago: University of Chicago Press, 2003.

———. *Before the Law: Humans and Other Animals in a Biopolitical Frame.* Chicago: University of Chicago Press, 2013.

Wood, David. "If a Cat Could Talk." *Aeon* (digital magazine), July 24, 2013. Accessed May 6, 2016. https://aeon.co/essays/the-uncanny-familiar-can -we-ever-really-know-a-cat.

Zahavi, Dan. "Empathy, Embodiment and Interpersonal Understanding: From Lipps to Schutz." *Inquiry: An Interdisciplinary Journal of Philosophy* 53, no. 3 (2010): 285–306.

———. *Husserl's Phenomenology.* Stanford, CA: Stanford University Press, 2003.

INDEX